Dare to Repair

Dare to Repair

A Do-It-Herself Guide to Fixing (Almost) Anything in the Home

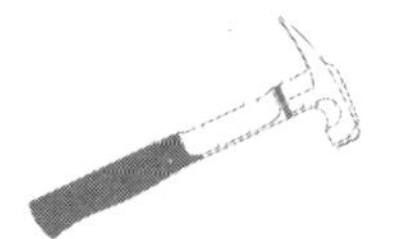

Julie Sussman &
Stephanie Glakas-Tenet

Illustrations by Yeorgos Lampathakis

HarperResource
An Imprint of HarperCollins*Publishers*

The OXO tools artwork has been provided with permission by OXO International.

HarperCollins books may be purchased for educational, business, or sales promotional use. For information please write: Special Markets Department, HarperCollins Publishers Inc., 10 East 53rd Street, New York, NY 10022.

FIRST EDITION

Library of Congress Cataloging-in-Publication Data

Sussman, Julie.

Dare to repair / Julie Sussman and Stephanie Glakas-Tenet ; illustrations by Yeorgos Lampathakis.—1st ed.

p. cm.

ISBN 0-06-095984-3

1. Dwellings—Maintenance and repair—Amateurs' manuals. 2. Do-it-yourself work. 3. Women. I. Glakas-Tenet, Stephanie. II. Title.

TH4817 .S87 2002

643'.7—dc21

2002027625

04 05 06 07 WBC/RRD 30 29 28 27 26 25 24 23 22 21

We believe God creates gentle souls—put on this earth to walk softly among others—as examples of how people behave in heaven. We have been blessed with two such people in our lives, our moms. We lovingly dedicate this book to them, Helene Johnson and Cleo Glakas.

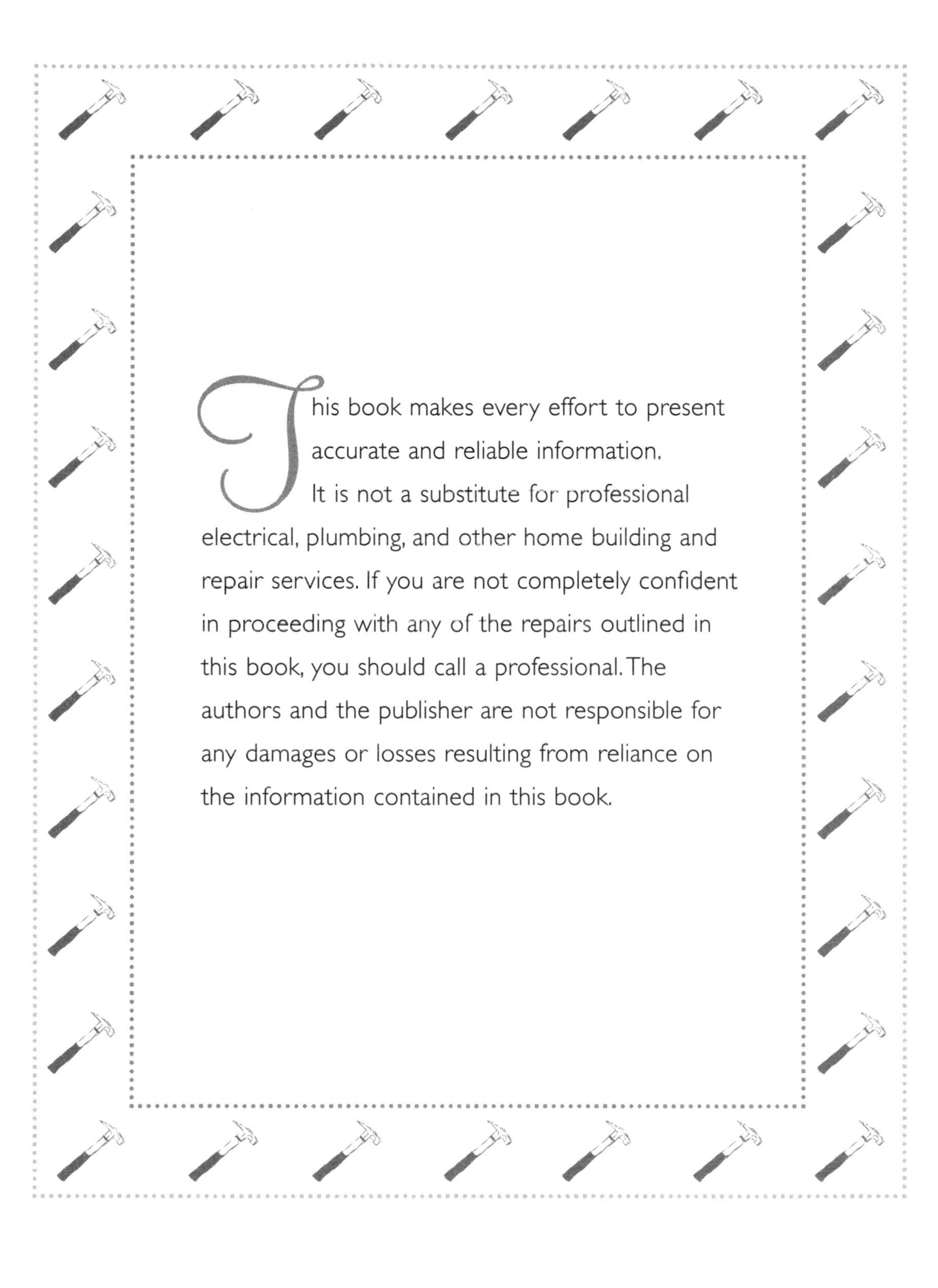

This book makes every effort to present accurate and reliable information. It is not a substitute for professional electrical, plumbing, and other home building and repair services. If you are not completely confident in proceeding with any of the repairs outlined in this book, you should call a professional. The authors and the publisher are not responsible for any damages or losses resulting from reliance on the information contained in this book.

Contents

Foreword

Four decades of adulthood has brought me face-to-face with more broken this or that than I care to remember. Growing up, my father was an ardent fixer-upper, my mother, a skilled seamstress. I learned from them how to use basic household tools and how to measure and cut accurately. I learned in time that if I couldn't fix something, it would be expensive to have an expert come in, so I would do it. Not only did I feel proud of myself for saving money but it also gave me great satisfaction to repair something. Too bad *Dare to Repair* wasn't around then!

I have passed some of these same skills on to my own children, who grew up in a throwaway society. If something broke, they often suggested that I throw it away and get a new one. But I encouraged them to fix it themselves to teach them something about saving money *and* landfill space. And when a repair job was successful, they (and I, too) would be amazed.

Today, as co-founder of Habitat for Humanity International, a Christian ministry founded in 1976 that provides opportunities for home ownership to low-income families, I often work alongside skilled as well as unskilled volunteers and future Habitat homeowners at various building sites around the world. Hence, my repair skills have now been expanded to construction skills! Interestingly, for some Habitat volunteers and homeowners, I have seen the reverse: People learning construction skills on Habitat building sites becoming more adept at repairs.

What I have learned at Habitat is that part of the American Dream has always been—and will always be—to own a home. The only thing that has changed is who's purchasing the house. It wasn't too long ago that the average home buyer was a man. Today, women are the fastest-growing segment of the housing market; in fact, 43 percent of Habitat for Humanity's homeowners are women heads of households. Through Habitat's Women Build program, thousands of women are being empowered and filled with joy as they discover, after a little instruction and encouragement, how easy it is to drive a nail, saw a board, and install sheet rock.

Each woman has her own story to tell of becoming a homeowner, and although the characters differ, the plot remains the same: a long journey down a difficult road sidetracked with disappointments, problems, and wrong turns. Yet, in the end they all say the struggle was well worth it because there is a universal joy in wiping your feet on your *own* welcome mat, jingling *your* house keys in your pocket, and writing *your* new address on an envelope.

The rewards of owning a home can last a lifetime and beyond—if you take care of it. Everyone knows it doesn't take long for a house to fall into disrepair without proper maintenance. Like constructing a home, *Dare to Repair* provides a you-can-do-it approach to fix-it jobs that have for too long been intimidating to most women—and even some men.

Dare to Repair will save you money, build your self-confidence to help yourself and your neighbors, and bring satisfaction to your heart and a smile to your face.

—Linda C. Fuller
Co-Founder, Habitat for Humanity International

Acknowledgments

We are honored to be able to thank the following people whose time, advice, and support enabled us to write this book.

To our agent, Liv Blumer (Karpfinger Agency), thank you for taking us under your wing and guiding us deftly through the publishing maze; our editor Kathy Huck, thank you for your grace under pressure, professionalism above and beyond the call of duty, and dedication to perfecting *Dare to Repair.* You were a delight to work with; Cathy Hemming (President and Publisher of General Books Group, HarperCollins) and Megan Newman (Editorial Director, HarperCollins), we are indebted to you for sharing our vision from the get go; Ginger McRae, our copy editor, thank you for keeping your eyes wide open for mistakes; the art staff of HarperCollins: Leah Carlson-Stanisic, Robin Bilardello, and Roberto De Vicq De Cumpfich, thank you for your creative strokes of genius. Yeorgos Lampathakis, our brilliant artist, we are grateful for your ability to give life to our words through your beautiful drawings; Athanasios Papapostolou, a gifted artist, thanks for pinch-hitting in the ninth inning; the wonderful people at OXO International: Gretchen Holt, Alex Lee, and Larry Witt, we thank you for supplying us with OXO tools to perform the repairs (we wouldn't work with anything else!) and for supporting our mission; our literary friends, Jeffery Deaver, Madelyn Warcholik, Sally Steenland, and Ralph Eubanks, thank you for generously sharing your knowledge; Kathy Lyons, we appreciated all the time you spent helping us and for not having the word *no* in your vocabulary; our accountant friend, Paul Wilner, and our attorney, Nina Graybill, special thanks for your professional advice and wisdom; The Roslyn Savings Bank, a heartfelt thanks for contributing to the cause; and Walter Mullins for being our guardian angel on earth. And a very special thank you to our dear friend Rosemary Ferrigno, for standing with us in the beginning, and later behind us, every step of the way.

We'd also like to single out some of the truly extraordinary people who provided us with technical information for *Dare to Repair*: Olivia Campos-Bergeron (OCB Consultants); Bob James (Northern Virginia Electric Cooperative); Frank Vecchio (American Standard Plumbing); Jo Rae Wagner (CTO, Inc., Mechanical and Plumbing Contractors); Bill Royston (All Pro Services, Inc.); Howard Brinegar, Jim Morrison, Larry Lauderdale, and Stan Smallwood (Home Depot); Andrew Trotta and Elizabeth Leland (Consumer Product Safety Commission); Ike Casey (Plumbing, Heating & Cooling Contractors National Association); Bill Whitehead (Plumbing and Draining Institute); Michael Harvey (Union Hardware, Inc.); Pete Reckendorf (Plumbing Parts Plus); David Stephens (Custom Building Products); Michael Clendenin (Electrical Safety Foundation International); Tim Kopp (Greenlee Textron); Lee Mathieson and Scott Norton (General Electric); Frank Stanonik (Gas Appliance Manufacturer's Association); Nate Hoffman (A Advantage Heating and Cooling); Dick Bray (Wood Floor Covering Association); Tom Palmer (DAP, Inc.); Beth Wortham (Purity Max); Jim Lake (National Fire Protection Association); Officer Sandy Redmond (Montgomery County Police Department); Bill Kistler (M.A.G., Inc.); Jay Graham (Automatic Specialties); Mark Bisbee (Liberty Carpet One); J. D. Grewell (J. D. Grewell & Associates, Inc.); Joe Tumulty (Tumulty Contracting Co.); Curt Forsgren (Aardvark Handyman Co.); Joe Lopes and Debbie Gooding (Printing Images); David Hollies and Ken Cherry (Home Connections, Inc.); Allison Gray (ServiceMagic.Com); and last but not least, Diane Hack Gould and Michael Gorski (D. H. Gould Company).

Personal Acknowledgments from Stephanie Glakas-Tenet

Without my co-author, Julie Sussman, *Dare to Repair* would never have been written. It was solely her vision and passion to improve the quality of women's lives that not only drove this project to completion but also steered me far beyond my own sights. Without my loving parents, John and Cleo Glakas, who provided me with my first toolboxes: one filled with hardware to repair things around the house, and the other filled with the tools—faith, forgiveness, and confidence—to fix anything in my life, I would never have known the stability of home. Without my husband, George, whose love and loyalty, instincts and intellect, humor and humility steady my course each day, I would never have known the mission of my life or the endless boundaries of my heart. Without my son, John Michael, with his passionate athleticism, sense of justice, perseverance, and fidelity to family and friends, I would never have known the wonders of motherhood. Without my incredible brothers, Nicky and Tommy, their wonderful families, and my dearest friends (the sisters I never had), I would never have known such immense joy, wisdom, and dependability. Their sense of direction has never failed me. Without my extended Agency family, whose patriotism, sacrifice, and discipline of duty have inspired me to stay the course, I would never have known the honor of serving my country. Having received all these blessings from a gracious God, I hope this book journeys on to be a blessing in the lives of others.

Personal Acknowledgments from Julie Sussman

I don't consider myself to be lucky. I know it took divine intervention to be graced with so many wonderful people in my life: my co-author, Stephanie Glakas-Tenet, who taught me the power of forgiveness and to see God in the details. She made writing this book a joy and a reality; my parents, Warren and Helene Johnson, whose teachings and examples of compassion, faith, and unconditional love inspire me to be a better person; my sisters and brother, Ann Walker, Mary Coyle, Chad Johnson, and Amy Marney, whose never-ending interest and support over the years motivated me to keep my dream alive. I am truly blessed to have such incredible siblings as lifelong friends. My children, Chad and Rebecca, whose love, laughter, and sweetness turned my black-and-white world into Technicolor with Surround Sound; and my husband and best friend, Jerry Sussman, whose love completes me. There is no kinder, gentler, loving man than he.

Dare to Repair

Introduction

Women have stood on the floor of the Senate, rocketed into space, sat in the chairs of the New York Stock Exchange, and climbed up corporate ladders. We've broken through glass ceilings—we just never learned how to fix them!

Why? Well, we have a theory: if a woman sees a bug in her home and knows a man is within screaming distance, she'll yell for him to kill the bug. But if a woman sees a bug and there's no man around, she'll smash it, stomp it, flush it, and do a victory dance. It's the same thing with home repairs—women can do them, we're just used to writing honey-do lists, waiting for dad to drop by, or calling the super.

We cooked up *Dare to Repair* out of sheer frustration. Things weren't getting fixed around our houses because our husbands were never home and we didn't have the money to hire contractors. We wanted to do the home repairs ourselves, but the do-it-yourself books on the market were written for tool-belted men, not for female repair rookies like us.

We knew we weren't alone in our situation: four times more women than men are heads of households. Women also make up 60 percent of all people living alone, have become the fastest-growing segment of home buyers, and are the driving force behind increased sales in hardware stores. Women want to be do-it-*herselfers*, but they haven't had the right home repair book. Their wait is over.

Dare to Repair fills the tremendous void left by other home repair books by providing basic repairs written in an easy-to-follow format with illustrations of women from all walks of life (finally, women who look like us and not the Barbies in our attics!). Each repair is introduced by an anecdote or pep talk intended to inspire you to read on.

The information, compiled from government agencies, national institutes, trade associations, manufacturers, and our own experience is accurate and up-to-date. We did every repair in *Dare to Repair* ourselves, so when we tell you it's easy, it is.

Neither of us was born with a silver hammer in her hand. We started doing home repairs out of financial necessity. You may have a different reason for needing to do the repairs yourself—maybe you're afraid to let a contractor into your home, or you can't take off from work to wait for one to arrive. Perhaps you need to spend your money elsewhere, or you just like a good challenge. No matter the reason, what's important is that you do whatever it takes to get started. If this means having a friend stand by to hand you tools along with moral support, do it. If you need to share the repair with a neighbor, do it. If it helps to keep a journal of your successes, do it.

If you start to panic and freeze with fear, stop, take a deep breath, and *get over yourself!* This isn't rocket science. And if this is the hardest thing you've ever done in your life, then *sister,* you haven't lived.

Think back to a time in your life when you challenged yourself, when you changed a flat tire, or asked your boss for a much deserved raise, or cared for your chicken-poxed children while your husband was overseas. Each of those victories counts as a notch in your success belt. Making repairs counts as a notch, too. Hey, it's the only time when adding notches to a belt is a *good* thing, so revel in it.

Our goal in writing this book is not only to teach you how to do home repairs but also to inspire you to pass your knowledge to others. When you learn a new repair, share it with someone else. Go to a neighbor and show her how to perform garage door safety tests. Visit your grandmother and secure the rugs in her home with double-sided tape. Install lever handles in your mom's home to make opening doors a little easier for her. Demonstrate to your sister how to use a fire extinguisher. Use your knowledge as a tool for improving your life as well as the lives of others.

Dare to raise the bar for what you can accomplish. *Dare* to pick up a wrench and tighten the toilet handle that's about to fall off. *Dare* to level the washing machine that's been rockin' and rollin' for months. Grab a screwdriver and *dare* to install a new smoke detector.

Dare to Repair!

Plumbing

Water Supply

- The Main Water Supply Valve: How to Find It and Shut It Off
- Restoring Hot Water to a Water Heater

Toilets

- Adjusting or Replacing a Toilet Handle
- Replacing a Toilet Seat
- Unclogging a Toilet
- Repairing a Running Toilet

Sinks and Bathtubs

- Clearing a Clogged Kitchen Sink Drain
- Unclogging a Bathroom Sink
- Retrieving Treasures from a Sink Trap
- Fixing a Bathroom Pop-Up Drain Stopper
- Unclogging a Bathtub Drain
- Replacing Sealant Around a Bathtub

Faucets

- Repairing a Leaking Faucet
- Unclogging a Sink Spray
- Replacing a Showerhead and Shower Arm

Seasonal Repairs

- Winterizing Plumbing
- Thawing a Frozen Water Pipe
- Clearing a Gutter and Downspout
- Bleeding a Hot Water Radiator

Plumbing problems, whether they involve sinks, toilets, or pipes, don't have to drain your wallet, energy, or time. The real difficulty often lies with getting over any fears (unfounded, of course) and motivating yourself to do the repair.

We've made this section simple so you'll want to take the plunge—we offer easy-to-follow instructions on only the most common plumbing problems. Heck, we even tell you when to throw in the towel and call a plumber.

For all you women who have never seen the inside of your toilet's tank and shudder at the thought of putting your hand into the water, relax. The tank's water is clean and the Tidy Bowl man doesn't bite.

Plumbing problems should never be fixed with a Band-Aid approach, nor should the repairs be put off, because they'll only get worse. Hell is *not* going to freeze over *girl*, so get going and fix whatever needs fixing!

Water Supply

The Main Water Supply Valve: How to Find It and Shut It Off

You need to know where the main water shutoff valve is located for emergencies and repairs.

As the wife of a military officer, Kathy knew that home is where the military sends you, so she was careful never to get too attached to a house. She also knew from experience that as soon as her husband boarded his ship, something in their base housing would break. Not wanting to join the ranks of a long waiting list for repairs, Kathy decided to start learning how to fix things herself. Her first order was locating the house's main water supply valve after the kitchen faucet broke off in her hand. It was rough seas at first, but now it's smooth sailing ahead.

Finding the Valve

Fresh water enters into your home through the main water supply line. The valve controlling the water flow through the line is typically found in the basement or utility room near the water meter, water heater, or on the front wall closest to the street. In older apartment buildings, the main water supply valve is located in the basement. However, in some new apartment buildings, main water supply valves are located on each floor in the utility room.

When you locate the main water supply valve, place an I.D. tag on it. If you want to make sure you've marked the correct valve, turn *on* a sink faucet and then shut *off* the main water supply valve. If the water from the sink stops, then you've found the right valve.

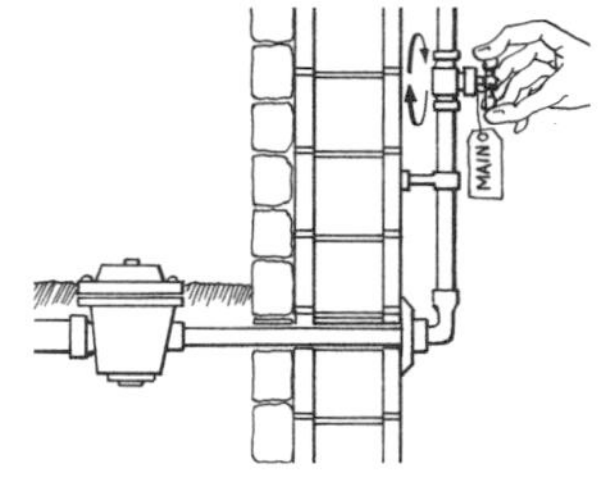

I.D. tag on main water supply valve

Determining the Valve Type

The two most common main water supply valves are gate and ball. The gate valve has a wheel-shaped knob that can be difficult to turn if it has been unused for a long time, or if it's dirty. A ball valve has a lever handle, which requires only a quarter turn to shut it *off*. Most new homes have the ball valve.

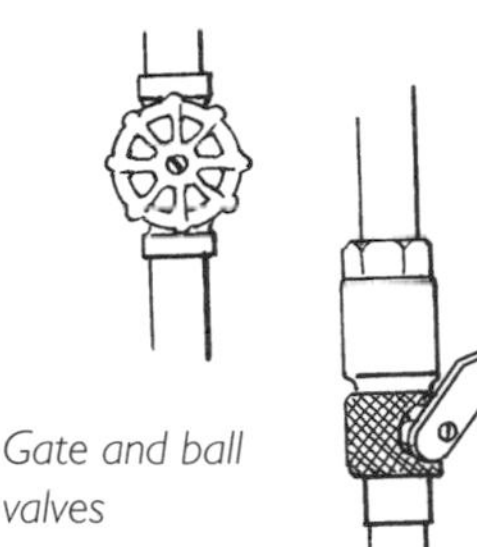

Gate and ball valves

We recommend you test the gate valve *prior* to an emergency, at least twice a year. And if it breaks, have it replaced with a ball valve.

Shutting Off the Valve

If you have a gate valve, turn it clockwise. If it's stuck, apply a lubricating spray, and use the adjustable wrench to turn it. For a ball valve, move the lever a quarter turn.

Loosening a gate valve

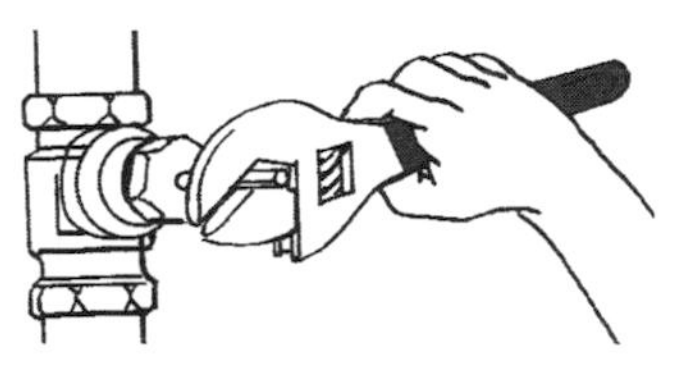

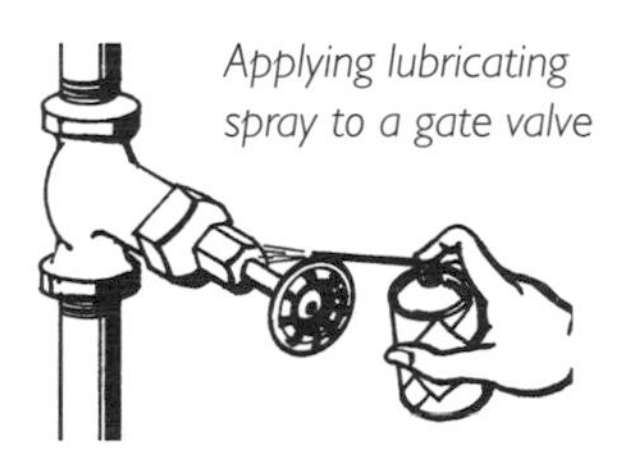

Applying lubricating spray to a gate valve

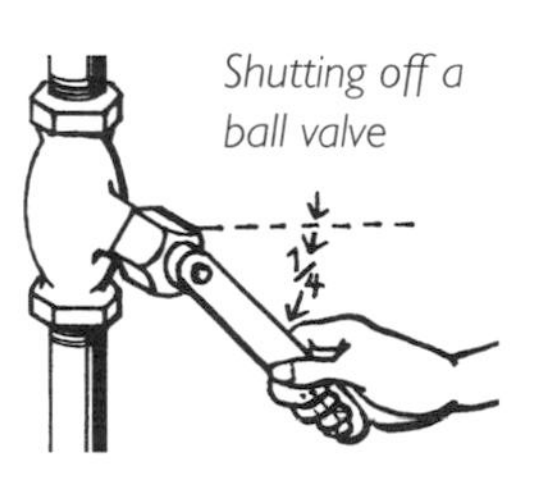

Shutting off a ball valve

Tools Needed

Adjustable wrench

Lubricating spray

Restoring Hot Water to a Water Heater

If your furnace and water heater are located in an open area where children play, use tape, string, chalk, or paint to establish a line of demarcation around the appliances and tell the children never to cross it.

Nancy came close once to hitting the high notes like Aretha, and it wasn't because she was happy. On a morning of a big meeting at work, she was in the shower when the hot water ran out, leaving her hair full of shampoo and only one leg shaved. Nancy raced to her gym not to exercise but to finish her shower!

Sound familiar? If you're not a member of the Polar Bear Club who happens to like icy cold water, a frigid shower can be un*bear*able. Chill out woman, the solution is easy.

Determining the Cause

The first thing you need to do is to find out why you're out of hot water. Possibly the shortage is due to heavy water usage during a brief time, an insufficient sized water tank, a leaking tank, or the water heater is not receiving power either because of a blown fuse or a tripped circuit breaker or its pilot light was extinguished (gas water heater).

Heavy Water Usage

If there are other people living in your home, stagger their showers.
If you live alone and you've depleted the hot water while singing "R-E-S-P-E-C-T," you need to wait about 20 minutes for the tank to heat up the water again.

Insufficient Sized Water Heater

Check the label on the water heater for the number of gallons it holds. Estimate 10 to 15 gallons per person per day for the right size water heater.

Leaking Tank

If the water heater is leaking, replace it immediately!

Lack of Power to a Water Heater

Before you restore power to a water heater, you need to determine which kind you have.

Every water heater will have at least one big sticker on it telling you the size of the tank and whether it's gas or electric. If for some reason you don't see any stickers, look on the water heater for a small metal label with serial numbers. The set of numbers starts with either an "E" for electric or a "G" for gas. And if you don't see the stickers or the metal label, then *sister,* we don't know what you're looking at!

Remove any gasoline containers, paint cans, paint thinner, or any other flammable materials from around the gas water heater. Either store them in a safe place away from the water heater or dispose of them responsibly.

Restoring Power to a Water Heater

If It's Electric

Check to see if your electricity is working. Go to the main service panel and look for a circuit breaker that has tripped or a fuse that has blown. If the problem is not with the fuse or circuit breaker, call a certified plumber, because the trouble lies within the water heater.

If It's Gas

There are two types of gas water heaters: one with a standing pilot and one with an electronic ignition. A standing pilot, which when operating correctly produces a blue flame, ignites the burner located at the base of the water heater. When a standing pilot goes out, it has to be manually lit.

An easy way to tell the difference between an electronic ignition

Some common reasons for a gas water heater not to work are excessive dirt, rust buildup, weak wiring, or a gust of wind that extinguished the pilot light.

The gas shut-off valve is typically located on the gas line parallel to the water heater. When the valve is in line with the gas line, the pipe is open; if it's perpendicular to the gas line, it's closed.

The average life expectancy for a water heater, gas or electric, is 12 to 15 years. You may get some signs before it dies, such as rust around the base of the heater, greatly increased flow of water, or unusually hot water. It's important to be attentive to this appliance, because when it goes, it will dump not only all of the water inside the tank but also any water coming into it from the main water supply.

and a standing pilot (if the flame is out) is to look for a cord. An electronic model has a cord that plugs into a 120-volt outlet or power source.

An electronic ignition is typically found in newer gas water heaters. If the burner goes out, you can't light it yourself. Before calling a certified plumber, contact the gas company to see if the gas supply was shut off.

Safety Rules Before Lighting the Gas Water Heater

First, sniff the area for gas. Natural gas is colorless and odorless, but the gas company adds a strong odor to it so you can detect a gas leak. If you smell gas, close the gas shut-off valve to the water heater. Do *not* turn on any lights or use your phone, because an electric spark can ignite the gas. Instead, leave your home immediately and contact the gas company from a neighbor's home or your cell phone. If you can't reach the gas company, call the fire department.

If the water heater is standing in water, don't touch it. Leave your home immediately and call the gas company.

If you don't smell gas, then proceed.

Notice that the tools we tell you to use do *not* include a wrench or pliers. Using a hand tool may cause a spark that could lead to an explosion, so use only your hands.

Lighting the Pilot of a Gas Water Heater

On the *top* of the control box is a large round knob called the gas control knob (a.k.a. gas cock). Turn it to the *off* position. On the *front* of the control box is a large round knob called the temperature dial. Turn it to the lowest possible setting. Now wait 10 minutes by your watch to allow any gas to escape. If you smell gas, stop and follow the safety rules above. If you don't smell gas, then continue.

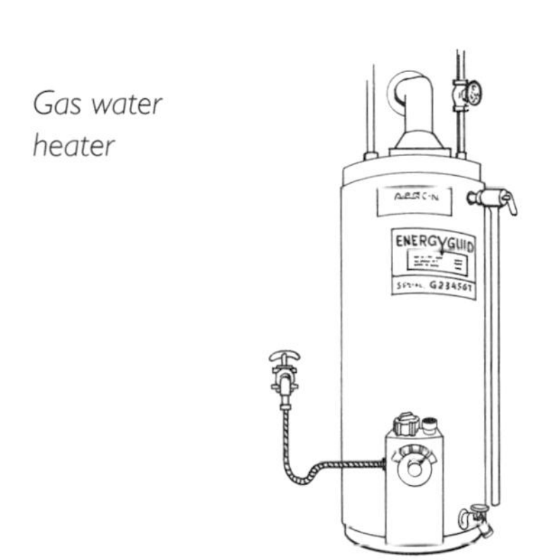
Gas water heater

Carefully remove the outer access panel. Before proceeding, read the instructions located either on or near the cover. If you don't see the instructions, contact the manufacturer.

Remove the inner access panel. Inside you'll see two metal tubes. The large tube is the pilot burner. Turn the gas control knob to the *pilot* position. Press down on the reset button (typically red), located next to the gas control knob. If your gas water heater does not have a

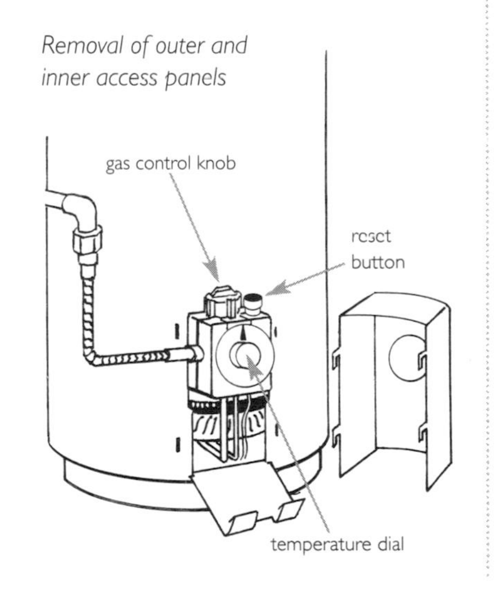

Removal of outer and inner access panels

Tools Needed

Watch  Barbecue lighter or long match 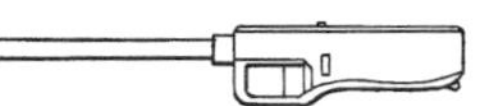

reset button, depress the gas control knob. While holding it down, light the burner with the barbecue lighter, keeping your face turned away for protection. Continue to hold down the reset button for 1 minute after the pilot is lit. Remove the lighter. Take your finger off the reset button, allowing it to pop back up.

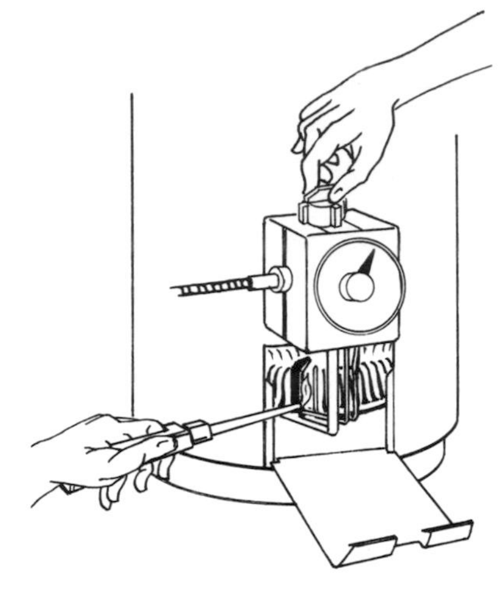

Lighting the pilot burner without the reset button

If the pilot remains lit, replace the inner and outer access panels. Turn the gas control knob to the *on* position. Turn the temperature dial to between 120°F and 130°F, or just above "Warm." When the burner ignites, you'll hear a whooshing sound. Turn the temperature dial back to 120°F.

If the pilot does not remain lit, repeat the process no more than twice. If it still does not light or remain lit, turn the gas control knob to the *off* position and contact the gas company or a certified plumber.

Toilets

Adjusting or Replacing a Toilet Handle

If your handle is jiggling more than Jell-O, there's no need to call in the cavalry or a super. All it takes is tightening the interior nut or replacing the handle.

Adjusting a Toilet Handle

Turn *off* the water at the toilet's shut-off valve (typically located behind the toilet on the wall or floor), and flush (probably twice) to remove as much water from the tank as possible. If you live in an older home and your toilet does not have its own shut-off valve, turn *off* the main water supply shut-off valve. Remove the top of the tank and gently place it in a spot where you won't trip over it.

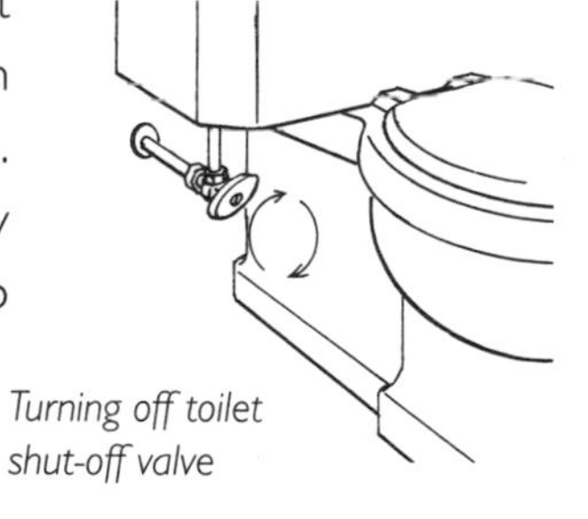

Turning off toilet shut-off valve

Tools Needed

Adjustable wrench  **Flathead screwdriver** 

Check the tightness of the nut located behind the handle on the inside of the tank. If it's loose, tighten it by using the adjustable wrench to turn the nut counterclockwise. (Note: This is the direction opposite from that you normally use to tighten a screw or nut.) Do not overtighten the nut, because you could crack the toilet. If the nut is not loose, proceed.

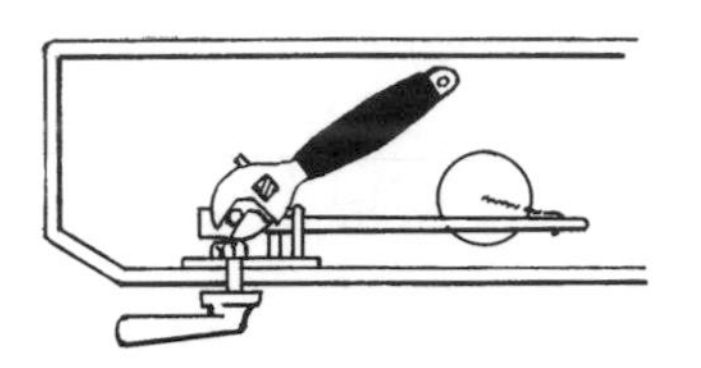

Turning handle nut with adjustable wrench

If Your Toilet Has a Lift Chain

You need to shorten the chain to tighten the handle. Simply detach the chain from the hook and move the hook further down the chain. Be careful not to make the chain too taut—you'll need some slack.

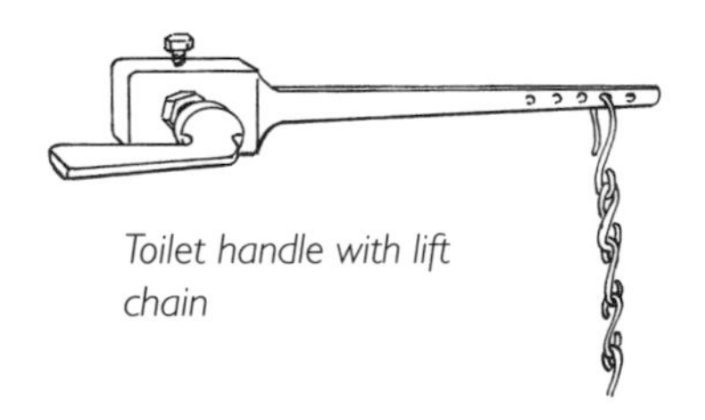

Toilet handle with lift chain

If Your Toilet Has a Lift Wire

You need to lift the tank ball higher by bending the wire upward or else adjust the guide arm with a screwdriver so that it moves freely. Turn *on* the water shut-off valve.

If this didn't fix the problem, then you'll need to replace the handle.

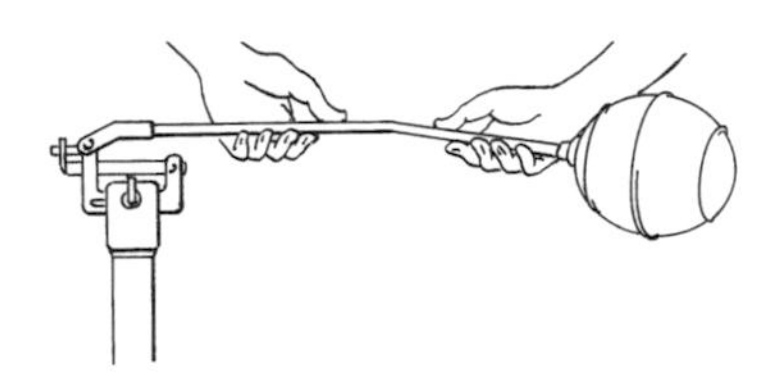

Bending lift wire

Replacing a Toilet Handle

Again, remove the top of the tank, turn *off* the water at the toilet's shut-off valve, and flush (probably twice) to remove as much water from the tank as possible. If you live in an older home and your toilet does not have its own shut-off valve, turn *off* the main water supply shut-off valve.

Removing an Old Handle

Remove the lift wire or chain from the trip lever (a.k.a. lift arm). Using an adjustable wrench, loosen the handle nut by turning it clockwise. (Note: This is the direction opposite from that you normally use to loosen a screw or nut.) If the nut won't budge, then apply lubricating spray, and use the adjustable wrench again. Remove the nut and pull the handle out along with the trip lever.

Before continuing, check the length of the original trip lever against that of the new lever. If the new lever is too long, you can cut off a piece of

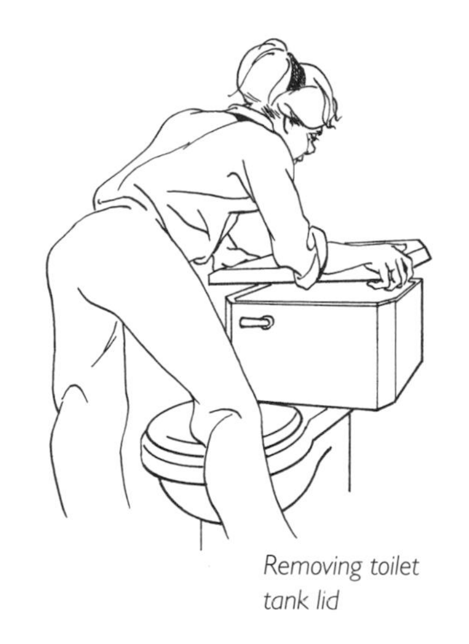

Removing toilet tank lid

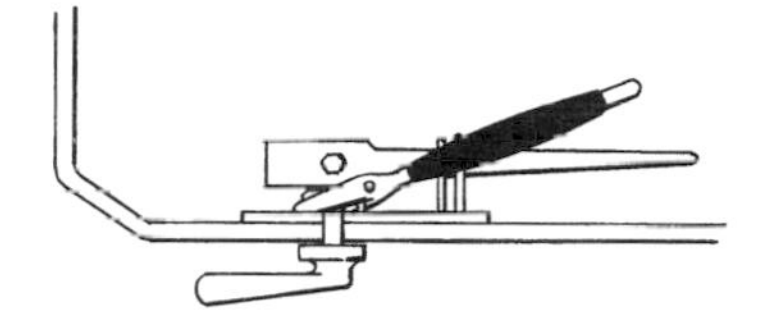

Loosening handle nut

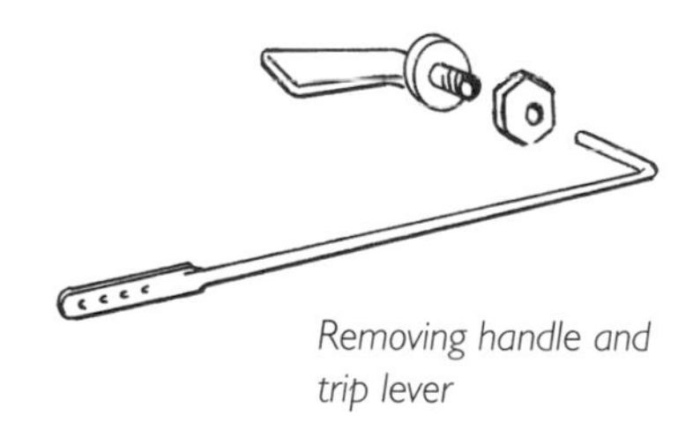

Removing handle and trip lever

Tools Needed

Adjustable wrench

Lubricating spray

Hacksaw or scissors

it with the hacksaw or scissors, being careful to cut where the manufacturer indicates.

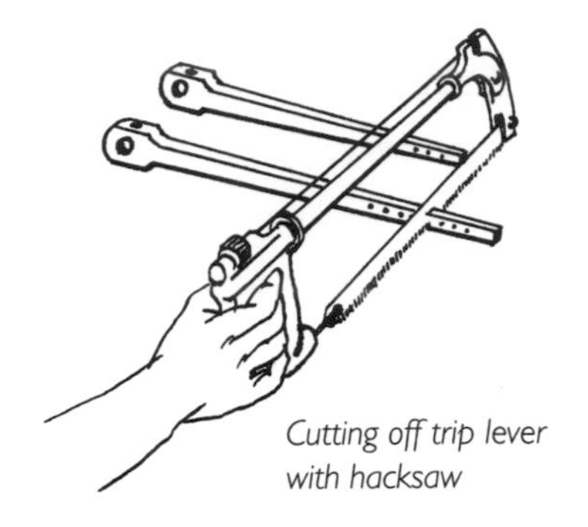
Cutting off trip lever with hacksaw

Installing a New Handle

Push the new trip lever through the hole and thread the nut (flat side facing the handle). Twist it counterclockwise, but don't tighten the nut completely just yet.

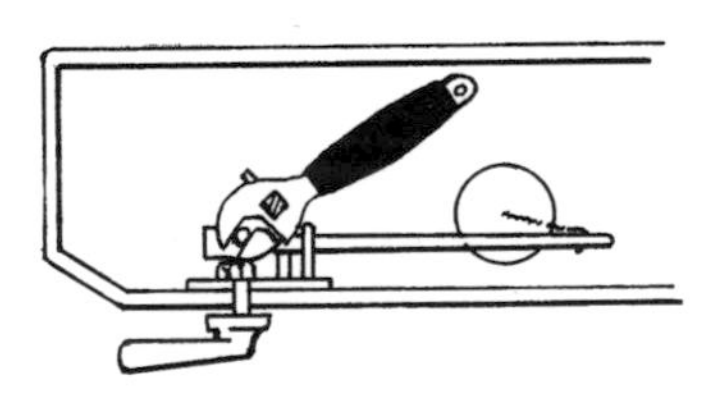
Turning handle nut with adjustable wrench

Connect the lift wire or chain back onto the trip lever, making certain to leave some slack. Tighten the nut using the adjustable wrench, but don't overtighten, because you could crack the toilet. Turn *on* the water shut-off valve.

Replacing a Toilet Seat

We once read in a home repair book that you should take your old toilet seat to the store when you want to buy a new one. The last thing this world needs is people walking up and down aisles with old toilet seats in their hands! There's only one place you should take your old toilet seat and that's to the nearest garbage can.

Whether you're purchasing a new toilet seat because the old one cracked, you want a color change, or you've just gotta try the puffy kind, it's important to make sure to match the shape and size of your toilet. If necessary, measure the distance between the two bolts for accuracy. We recommend you purchase toilet seats with plastic rather than metal bolts and nuts, because they won't rust and therefore are easier to remove.

Removing an old seat

The bolts attaching the seat to the toilet are either exposed or covered by plastic caps. If your toilet has plastic caps, use the screwdriver to lift them up.

Tools Needed

Flathead screwdriver 

Adjustable wrench 

Lubricating spray

New toilet seat 

While holding on to the nut (located underneath the bolt) with the adjustable wrench, turn the bolt counterclockwise with your fingers or screwdriver. Remove the bolt and nut and repeat the process on the other side. If the nuts and bolts are metal and you're having difficulty removing them, apply lubricating spray before using the adjustable wrench again.

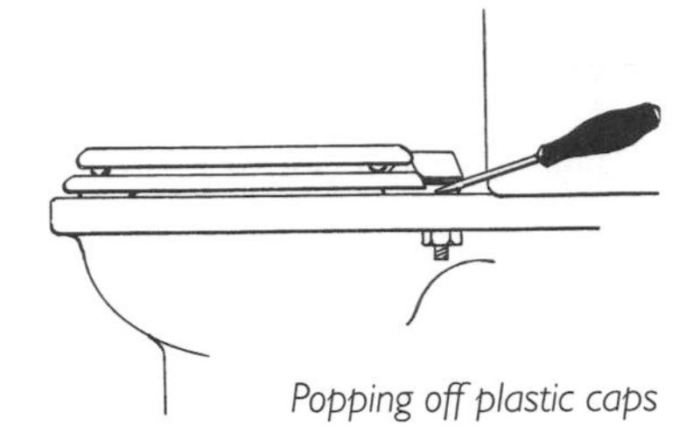
Popping off plastic caps

Take the toilet seat off and throw it away. Clean the area.

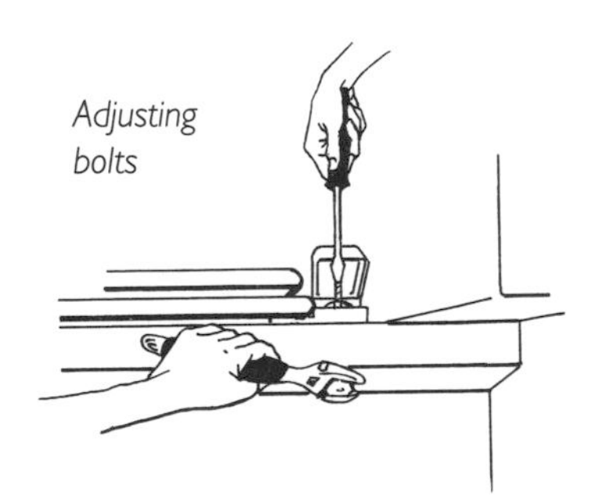
Adjusting bolts

Attaching a New Seat

Place the new toilet seat on top of the toilet, positioning it so that the mounts are directly over the holes. Put a bolt through the mount and thread the nut on the bottom of the bolt.

While using the adjustable wrench to hold the nut in place, turn the bolt clockwise with your fingers or screwdriver. Use the adjustable wrench to tighten the nut. Repeat the process on the other side. If you have plastic covers, snap them shut.

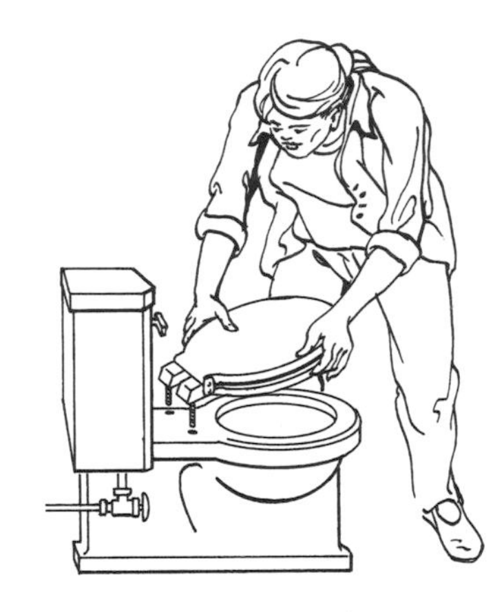
Installing new toilet seat

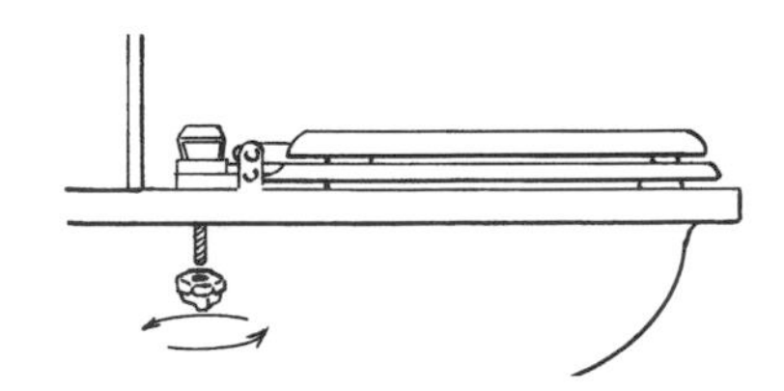
Threading nut and bolt

Unclogging a Toilet

Pat's toilet was clogged and no matter how hard she tried to fix it, nothing worked. An expensive visit from a plumber revealed the cause of the problem—dental floss. The plumber explained to her that a lot of people dispose of their floss in the toilet, not realizing that it gets stuck in the rough insides of the sewer pipes. "It's a big price to pay for 'porcelain'-white teeth," he joked. "You 'crack' me up," replied Pat dryly, as she viewed his backside in the mirror.

Never try to unclog a toilet with drain cleaner, because it can damage the pipes.

If you never seem to have bathroom trash, it's a good bet that someone is using the toilet as a trash can. Napkins, paper towels, feminine products (that aren't flushable), and dental floss (as Pat learned) are the most common culprits that clog a toilet.

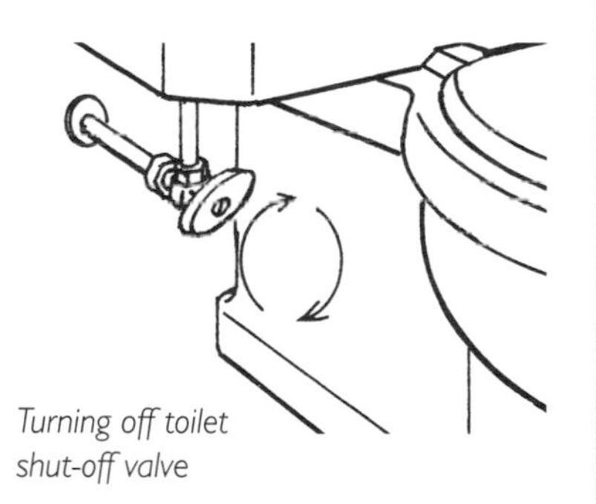

Turning off toilet shut-off valve

Tools Needed

Toilet plunger 

Bucket 

Closet auger (weird name, but it's actually for the toilet) 

Turning Off the Water

Turn *off* the water at the toilet's shut-off valve (typically located behind the toilet on the wall or floor). If you live in an older home and do not have a shut-off valve, turn *off* the main water supply shut-off valve.

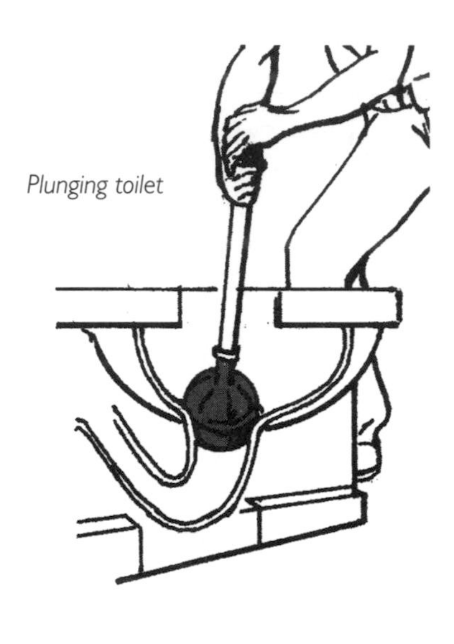

Plunging toilet

Using a Plunger

If the bowl is less than half full, use the bucket to add more water. Place the plunger in the toilet over the drain hole and rapidly pump at least a dozen times without breaking the suction.

Pour a bucket of water into the bowl to move the obstruction through the drain hole. If the water does not go down the drain hole, use the plunger again.

A closet auger can be purchased at a plumbing supply or hardware store.

Using a Closet Auger

If the water is still not draining, place the corkscrew end of the auger into the drain hole, maneuvering the rubber tubing completely down the metal spring to avoid marring the bowl. While inserting the auger, apply pressure, turning the handle clockwise until the entire spring has been fed through and reaches the obstruction.

Slowly pull the spring back out, turning the handle clockwise. If this does not free the obstruction, use the auger again (several times). Once the toilet has been unclogged, turn *on* the water shut-off valve. If the obstruction didn't budge, call a certified plumber.

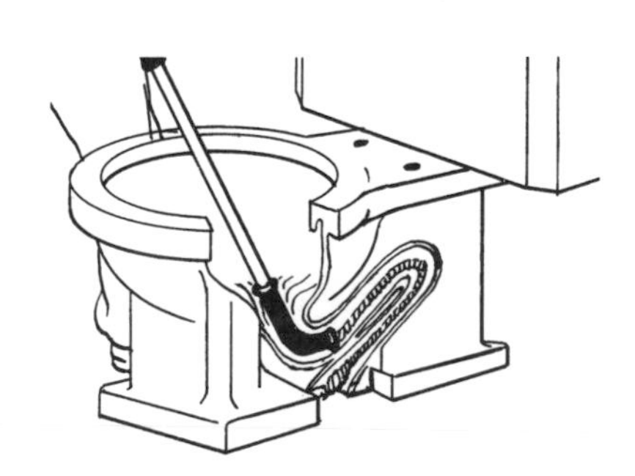

Inserting closet auger

Repairing a Running Toilet

When Retta was pregnant, she made regular nighttime visits to the bathroom. It was after one of those trips that the dam broke—not hers, the toilet's! While Retta was asleep, the toilet continually filled with water, so that by morning, water had poured through the bathroom floor into the kitchen below, across the hallway, and down the stairs. There was so much water that Retta swore she saw animals passing two by two.

Our guess is that Retta knew her toilet was running *before* the accident, which meant that it could have been avoided with a quick fix.

A toilet is "running" when you hear constant water activity in the toilet long after usage. Not only is a running toilet noisy but it also wastes a lot of water, and therefore should be fixed immediately.

One of the reasons a toilet runs is that the tank's water level is too high. If the water level in the tank is above the overflow tube, the water will run off into the tube, which then sends it into the bowl. (Note: If the water level in the tank is too low, the bowl will not fill enough

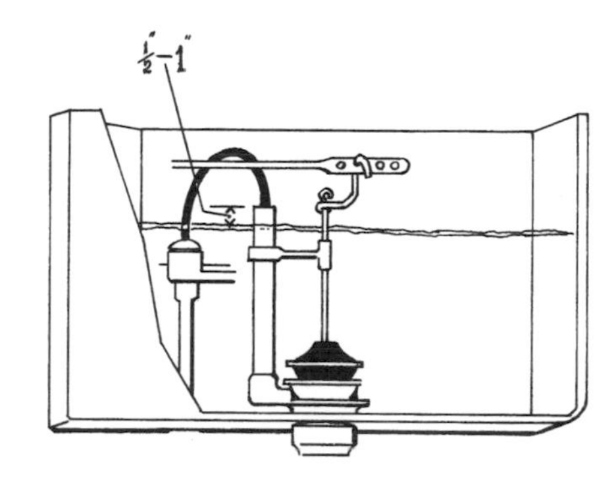

Identifying tank's water level

for an adequate flush.) The tank's water should always be ½ inch to 1 inch below the top of the overflow tube.

Your toilet will have one of three types of flushing mechanisms: 1) float arm; 2) float cup; or 3) metered fill valve. When the tank's cover is off, refer to the illustrations to know which kind of mechanism is in your tank.

Adjusting the Tank's Water Level

For a Float Arm

Slightly bend the metal arm downward so that the float ball is lower than before. To lower or raise a plastic float arm, adjust the knob at the ballcock. Flush the toilet to check the level and repeat the process, if necessary.

If the toilet bowl does not have enough water for an adequate flush, slightly bend the metal arm upward and flush to check the level in the tank and bowl.

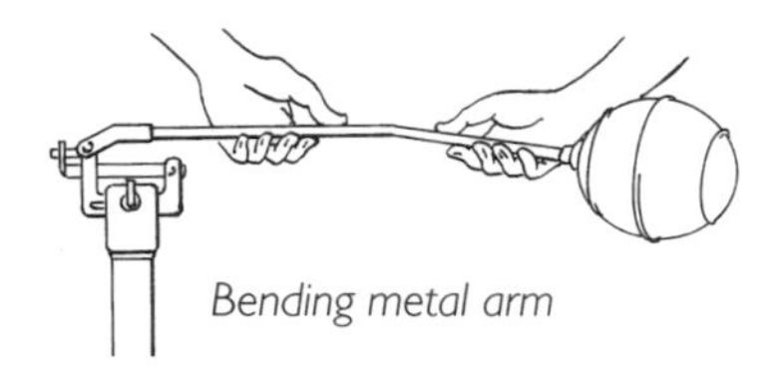

Bending metal arm

For a Float Cup

Adjusting a float cup model requires you to squeeze the metal clip to slide it up to raise the water level, and down to lower the water level. Flush the toilet to check the level and repeat the process, if necessary.

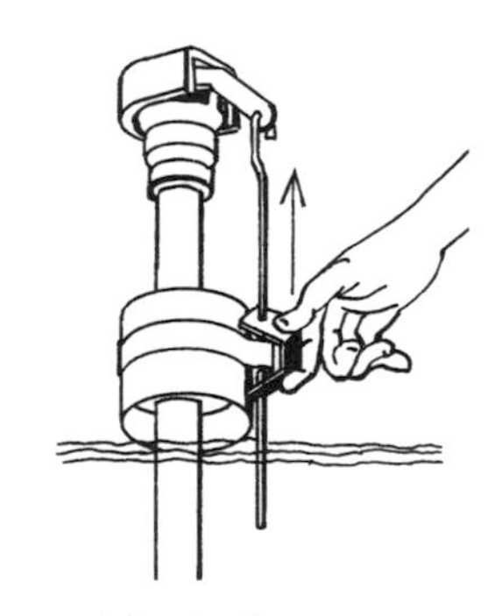

Adjusting float cup

Tool Needed

Flathead screwdriver 

For a Metered Fill Valve

The metered fill valve is the oldest type of flush system. Use a flathead screwdriver to turn the screw (located on the fill valve) clockwise to raise the water level and counterclockwise to lower it, moving the screw a half turn each time. Flush the toilet to check the water level and repeat the process if necessary.

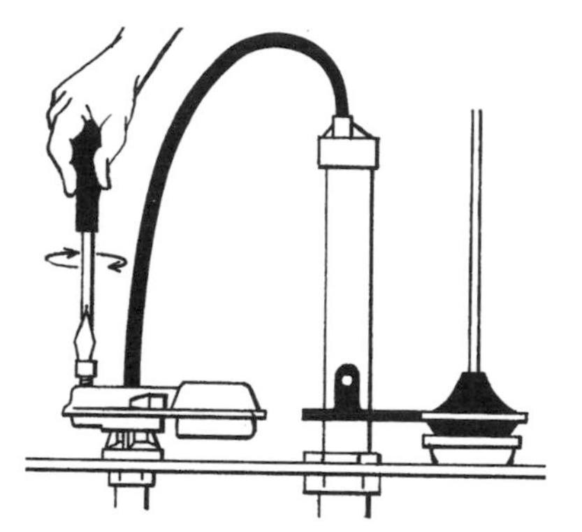

Adjusting metered fill valve

If any of these adjustments fail, you may need to replace the toilet flapper or the entire fill valve (see following repair).

Adjusting or Replacing a Toilet Flapper

If the toilet is still running after adjustments to the tank's water level, and you find yourself jiggling the toilet handle, then the problem lies with the toilet flapper (also called stopper).

Adjusting a Flapper

Turn *off* the water at the toilet's shut-off valve (typically located behind the toilet on the wall or floor). If you live in an older home and you do not have a shut-off valve for your toilet, turn *off* the main water supply shut-off valve. Remove the tank's lid and place it where you won't trip

Tools Needed

Sponge 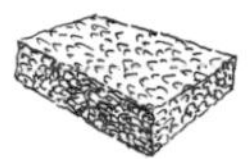 Flathead screwdriver 

over it. Flush the toilet (probably twice) to remove as much of the water in the tank as possible, using the sponge to absorb any remaining water. When you flush the toilet, notice if the flapper lands directly onto the flush valve opening; if it doesn't, you'll need to realign the flapper to fix the problem.

If your tank has a lift wire instead of a lift chain, the wire may be too close to the overflow tube, causing it to impede flushing. First try bending the lift wire slightly, away from the guide arm. If unsuccessful, use the screwdriver to loosen the screw on the guide arm to adjust the lift wire. Turn *on* the water to the toilet and flush. If the flapper still doesn't work, then you'll need to replace it.

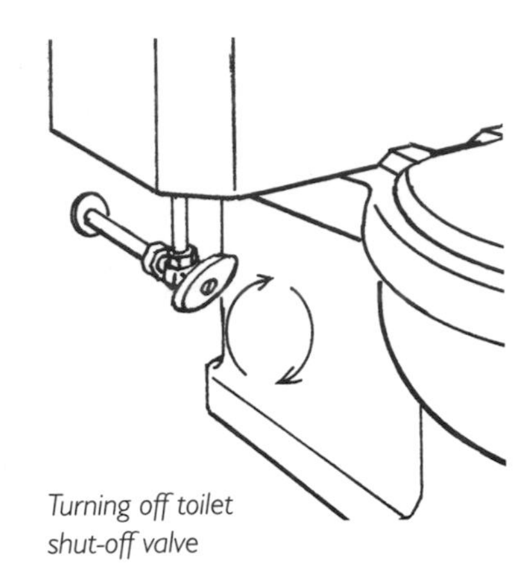

Turning off toilet shut-off valve

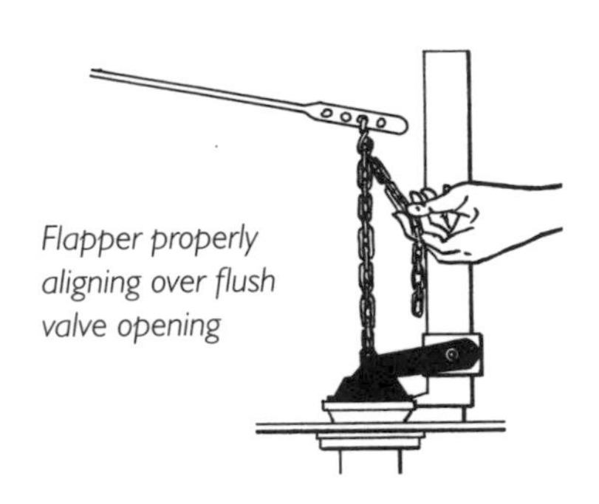

Flapper properly aligning over flush valve opening

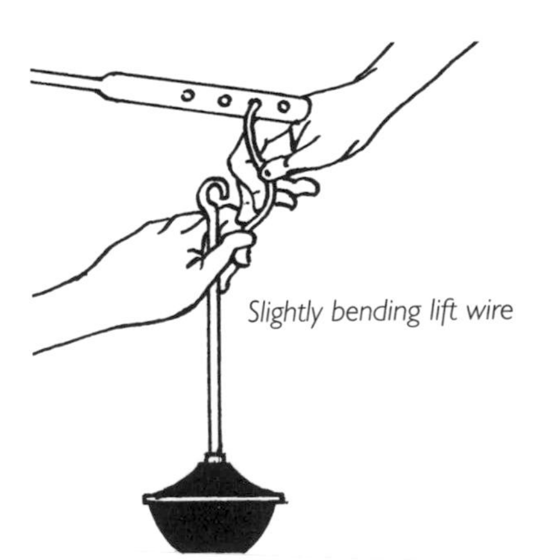

Slightly bending lift wire

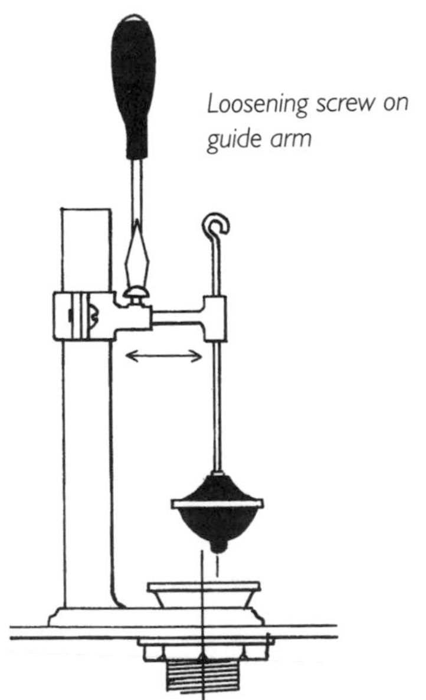

Loosening screw on guide arm

Replacing a Flapper

If the flapper is old and no longer functioning, replace it with one that comes with a lift chain and collar attached to it.

Turn *off* the water at the toilet's shut-off valve. If you live in an older home and you do not have a shut-off valve for your toilet, turn *off* the main water supply shut-off valve. Remove the tank's lid and place it where you won't trip over it. Flush the toilet (probably twice) to remove as much of the water as possible, using the sponge to absorb any remaining water.

Remove the old flapper along with its lift wire and guide arm. Slide the new flapper's collar, with the flapper attached to it, down the overflow pipe. Attach the chain to the trip lever by hooking it into one of the holes, allowing some slack.

Turn the water *on* at the shut-off valve and wait for the tank to fill before flushing. If the tank's water doesn't drain completely, make adjustments to the chain and trip lever.

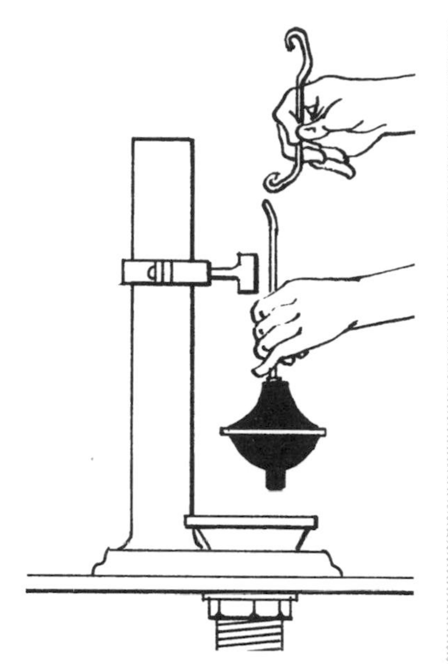

Removing old flapper, lift wire, and guide arm

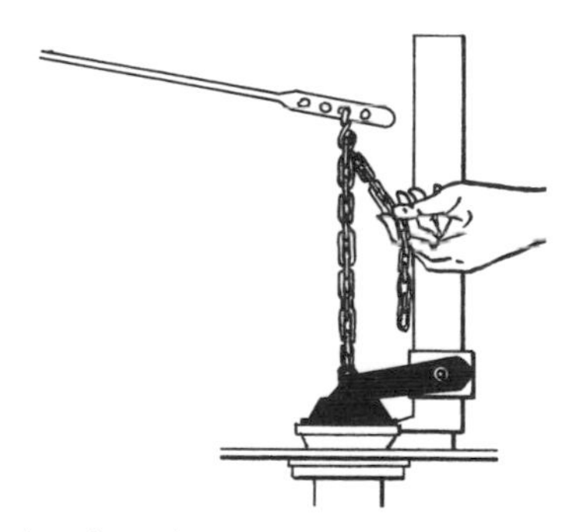

Attaching chain to trip lever

Be careful—the pigmentation from an old flapper may come off on your hands.

Tools Needed

Sponge 

New flapper 

After you've adjusted the tank's water level and replaced the flapper and the toilet still isn't working, it's time to install a new toilet fill valve (see following repair).

Replacing a Toilet Fill Valve

This repair sounds more complicated than it is, probably because you're not familiar with the wacky names of some of the parts. It can be challenging only if the nuts are extremely rusted and/or there's little space to maneuver around the toilet, so be patient. You *can* do it!

We feel obliged to warn you that before you begin this repair it pays to clean your toilet and its surrounding area, because you'll be very up close and personal with it, especially if the toilet is in tight quarters.

Purchasing a New Fill Valve

If the flushing mechanism in your tank looks different from our illustration, you probably have an older type, called a ballcock. We recommend you replace it with a newer type called a *fill valve*.

A new fill valve will come with these parts: shank washer, cone washer, O-ring, coupling nut, and angle adapter. It's important to read the instructions, because some manufacturers will attach

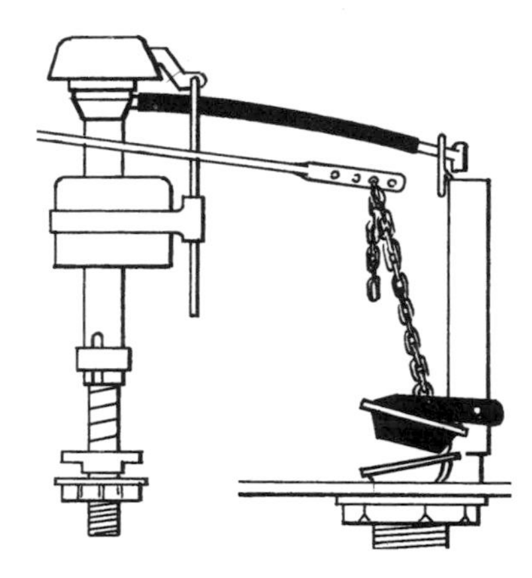

Flushing mechanism with fill valve

Tools Needed

Sponge 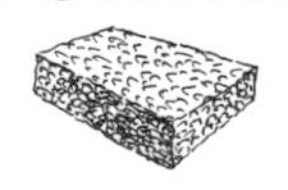 Towel  Lubricating spray 

Slip-joint pliers or plumber's wrench  Adjustable wrench 

Scouring pad  New fill valve 

all the parts, requiring you to separate them. For example, sometimes a cone washer comes attached to a shank washer, and if you've never done this repair before, you'd assume a part is missing.

Don't overtighten plastic nuts, because you can crack the toilet.

Removing the Old Fill Valve

Turn *off* the water at the toilet's shut-off valve (typically located behind the toilet on the wall or floor). If you live in an older home and you do not have a shut-off valve for your toilet, turn *off* the main water supply shut-off valve.

Remove the tank's lid and place it where you won't trip over it. Flush the toilet (probably twice) to remove as much of the water in the tank as possible, using the sponge to absorb any remaining water.

The tank needs to be as empty as possible, because when the fill valve is removed, there will be an opening in the tank where water can pour out. If water is still coming into the tank after you've turned *off* the shut-off valve, you probably didn't close the valve completely. If all else fails, turn *off* the main water supply.

Place a towel on the floor on the same side as the toilet handle. First, you're going to remove the coupling nut and the locknut, both located on the supply tube. If the nuts are rusted, apply lubricating spray. Using the slip-joint pliers, loosen the coupling nut by turning it clockwise (opposite from the usual direction) and slide it off the tube.

You'll need both hands to remove the locknut (located on the exterior of the tank, directly below the fill valve). Secure the slip-joint pliers to the base of the fill valve, or grip it tightly with one hand. With the other hand, place the adjustable wrench on the locknut and turn it clockwise to remove. Take out the refill tube before pulling out the old fill valve.

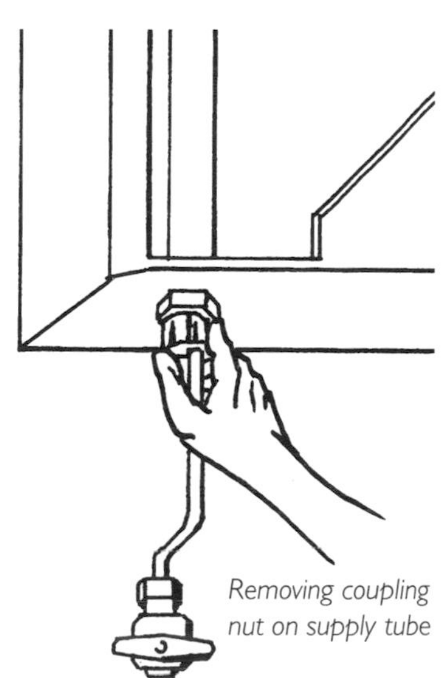

Removing coupling nut on supply tube

Installing a New Fill Valve

Compare the height of the old fill valve with the new one, and make any necessary adjustments by twisting the base of the new fill valve to either extend or shorten it. Clean the area around the hole in the bottom of the tank to provide better suction for the new shank washer.

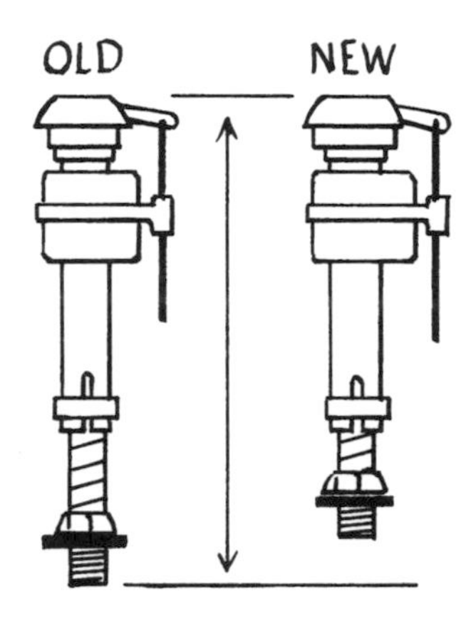

Comparing old and new fill valves

A new fill valve is typically made of plastic. Older models are made of copper or brass.

Remove the cone washer from the shank washer, if they're attached. Place the shank washer on the fill valve with the flat side up, facing the fill valve. Insert the fill valve into the toilet. While holding the fill valve with one hand, fit the new locknut on the supply tube, turning it counterclockwise with your fingers to tighten it. Use the slip-joint pliers (or your hand) to hold the fill valve as you tighten the locknut with the adjustable wrench.

Refer to the manufacturer's instructions to determine whether your supply tube requires the use of the existing nuts and washers versus new ones, and what is the exact order in which they should be installed. The most common type of supply tube, metal/copper tubing, requires the cone washer to be inserted into the new coupling nut before you screw it onto the supply tube with the slip-joint pliers.

Connect one end of the refill tube (typically black) to the fill valve. Insert the other end of the refill tube onto the angle adapter and clip it to the edge of the overfill tube.

Pinch the spring clip on the float cup to slide it up or down. Remember, the tank's water level should be ½ inch to 1 inch below the overfill tube. Turn *on* the water to the toilet and flush. Look to see that the tank's water is at the correct level; if not, adjust the float cup. Replace the tank's lid.

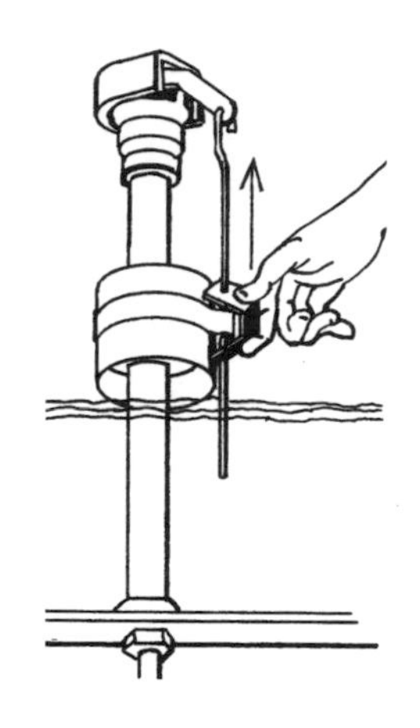

Adjusting float cup

Sinks and Bathtubs

Clearing a Clogged Kitchen Sink Drain

'Twas Christmas night when Katy's kitchen sink and garbage disposal decided that 20 pounds of potato peel was enough and quit cold turkey! Later, while Katy and her husband were washing the dishes (in the bathtub), they talked of how they couldn't afford a plumber and decided that the more "handy" of the two would have to fix it. With the right tools and determination, Katy tackled the job successfully. Total cost: $4.49. Newfound confidence: Priceless.

Never use drain cleaner to unclog a sink, because it can damage pipes, gaskets, and garbage disposals.

We know that sometimes sinks (bathroom and kitchen) will clog no matter how good you are at maintaining them, and that's why it's important to keep plungers handy.

Notice that we put an s on the word *plunger.* We recommend you have two plungers in your home—one for sinks, one for toilets, and never the two shall meet. Get our drift?

Tools Needed

Rubber gloves 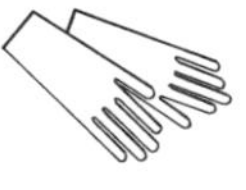

Petroleum jelly 

Plunger 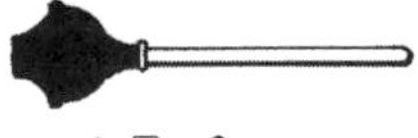

Sink stopper 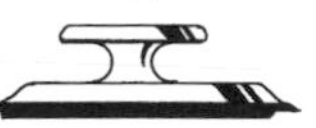

Small C-clamp, or rag and helpful friend 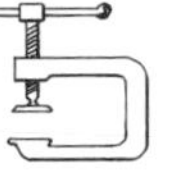

First remove the sink strainer and, while wearing rubber gloves, clean out any matter that may be in the drain hole. Remove the gloves.

Now here's where there's a fork in the road (*hee hee*): the method of repair you'll use depends on whether you have a single or double sink, and if you have a garbage disposal.

Single Sink Without a Garbage Disposal

Apply some petroleum jelly to the rim of the plunger and place it directly over the drain hole. If the water in the sink does not cover the plunger cup, add water. Pump the plunger vigorously, at least a dozen times. If the water hasn't dissipated, continue pumping at least a dozen times with each try.

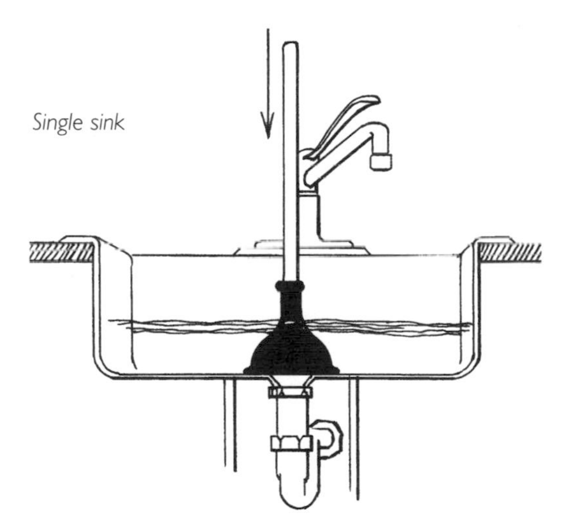

Double Sink Without a Garbage Disposal

Insert the sink stopper into the drain hole on the side that's *not* clogged. It's important to do this, because otherwise the water will go from one sink to the other instead of forcing the debris through the drain. Apply some petroleum jelly to the rim of the plunger and place it directly over the drain hole. If the water in the sink does not cover the plunger cup, add water. Pump the plunger vigorously, at least a dozen times. If the water hasn't dissipated, continue pumping at least a dozen times with each try.

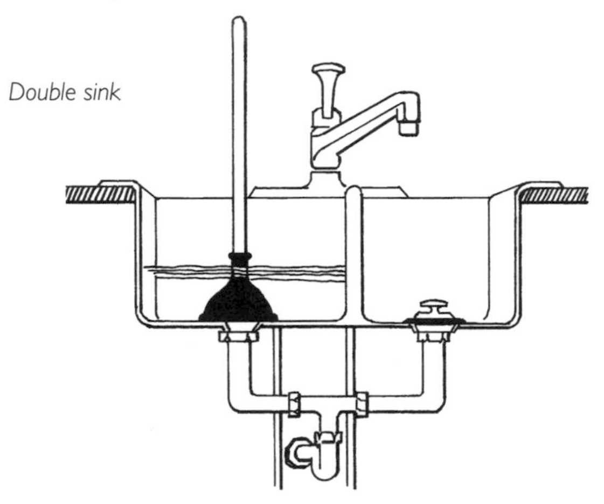

Double sink

Single or Double Sink with a Garbage Disposal

For a double sink, insert the sink stopper into the drain hole of the sink that's *not* clogged.

If your sink has an air gap, you'll need to block the water that would otherwise come through it while you're plunging. Why? you ask. Because that baby can shoot water pretty far! You can block the water either by using a small C-clamp to close off the hose that connects the air gap to the garbage disposal, or by enlisting a helpful friend to hold a rag over the air gap while you use the plunger.

If you're doing this by yourself, find the hose that connects the air gap to the garbage disposal, located underneath the sink. It will be the hose that's ribbed and closest to the air gap. Attach the C-clamp to the hose to block any water from entering it.

An air gap is a chrome-fitting dome, about 3 inches tall, that sits next to the faucet. Its purpose is to prevent back slippage of contaminated water into the dishwasher.

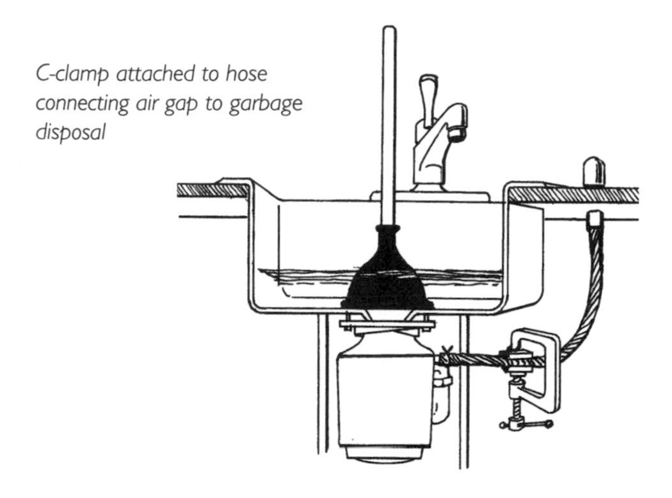

Apply some petroleum jelly to the rim of the plunger and place it directly over the drain hole. If the water in the sink does not cover the plunger cup, add water. Pump the plunger vigorously, at least a dozen times, until the water goes down the drain. Remove the C-clamp.

If the rapid plunging doesn't work, you'll have to remove the sink trap (see "Retrieving Treasures from a Sink Trap," page 36).

Unclogging a Bathroom Sink

Jeryl, Rosemary, and Patti owned a full-service beauty salon. In the beginning, a clogged sink meant that business was good. But as their clientele grew, regular visits from the plumber caused *shear* havoc with their profits. So, the barber babes took action by enrolling in a plumbing class. Now their money goes into their pockets instead of down the drain.

Never use drain cleaner to unclog a sink, because it can damage pipes, gaskets, and garbage disposals.

There are a lot of people (including the barber babes) who would like a garbage disposal specifically designed for a bathroom sink. But until that day comes, there are simple ways to unclog the sink.

Before you start, notice that the sink has an overflow hole (a kitchen sink does not) that allows excess water to drain into it. To ensure that the water in the sink goes down the drain and not into the overflow hole, you need to close it off before using the plunger.

If your sink has a drain stopper, remove it. Insert the rag into the overflow hole located in the side of the sink opposite the faucet (lean over and look into the sink, or feel for it with your finger).

Tools Needed

Rag  Plunger 

Place the rim of the plunger directly over the drain hole. If the water in the sink does not cover the cup of the plunger, add water. Pump the plunger rapidly, at least a dozen times with each try. Repeat if necessary. Remove the rag and replace the drain stopper.

If plunging doesn't work, then remove the sink trap (see "Retrieving Treasures from a Sink Trap," page 36).

Rag inserted in overflow hole while plunging

Retrieving Treasures from a Sink Trap

A lot of women (including us) own jewelry boxes full of pair-less earrings. Maybe it's because we think the matching ones are still in the sink drain and there's hope for a search and rescue mission. Unfortunately, the missing jewelry is forever gone to the big, bad sewer. Rats!

If a piece of jewelry, or another object, falls into the sink, immediately turn *off* the water and cover the drain hole with a sink stopper (or anything else you can find), because any water that enters the drain will force the jewelry farther into the drainpipe.

Underneath the sink you'll notice a curved pipe called a trap. The trap acts as a barrier to prevent sewer gases from leaking back into the house, as well as to keep objects that might clog the main drain from entering it. If your home is new, the trap is probably made of PVC (plastic) piping; if you have an older home and the pipes have never been replaced, the trap may be a combination of copper and chrome. Both types have two slip nuts that need to be either removed or loosened and slipped aside so that the trap can be dismantled.

You may also need to remove a sink trap if it's leaking. Take the original trap to the hardware store for a replacement if you can—be warned that if your chrome/copper sink trap is old and rusted, it may crumble when you remove it.

Removing an Object from the Trap

Put the bucket underneath the trap. Wrap the head of the slip-joint pliers with masking tape to prevent marring. Place the slip-joint pliers around the slip nut and turn it counterclockwise until it is loosened. Repeat for the second nut. Apply lubricating spray if the nuts are difficult to move.

For chrome/copper piping, slide the slip nuts down and remove the washers. For PVC piping, remove the slip nuts and washers. Disconnect the trap, being careful not to spill the water collected in the bottom of it. Empty the contents into the bucket. If the fallen jewelry is not there, it's possible that it's stuck in the drainpipe.

Remove the sink stopper, allowing any water in the sink to go through the drainpipe into the bucket, hopefully dislodging the jewelry. If you still don't find the gem, then it entered the main drain and is gone forever.

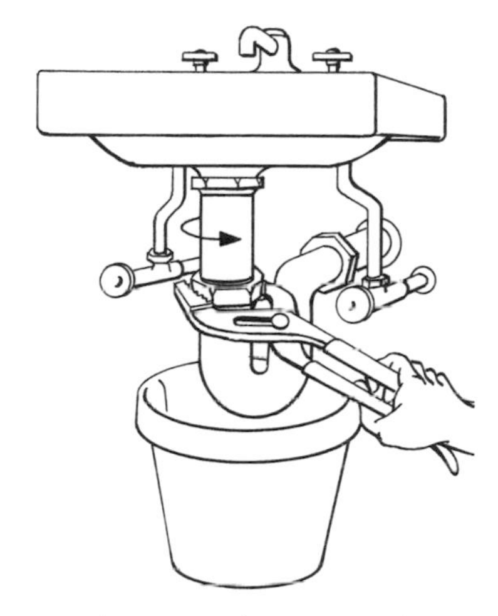

Loosening sink trap assembly

Emptying contents into bucket

Tools Needed

Bucket 

Masking tape 

Lubricating spray 

Slip-joint pliers or plumber's wrench 

While you're at it, check for wear and tear on the washers and trap and replace if necessary. Take the old trap and washers to the plumbing or hardware store for a matching replacement.

Attaching a New/Old Trap

To attach a new/old trap, first slide the slip nut, and then the washer, onto the tailpiece. Connect the straight end of the trap with the tail-piece. Slip the washer over the joint (i.e., where the two pipes meet) and move the slip nut on top of the washer. Tighten *loosely* with your hand, making sure that the slip nut is going on straight.

Place the slip nut, and then the washer, onto the trap arm. Connect the bend of the trap with the trap arm. Slip the washer over the joint and move the slip nut on top of the washer. Tighten *loosely* with your hand, making sure that the slip nut is going on straight.

Use the slip-joint pliers to tighten both slip nuts. Don't take the bucket away until you turn the water *on* to check for leaks. If there is a leak, re-tighten the slip nuts.

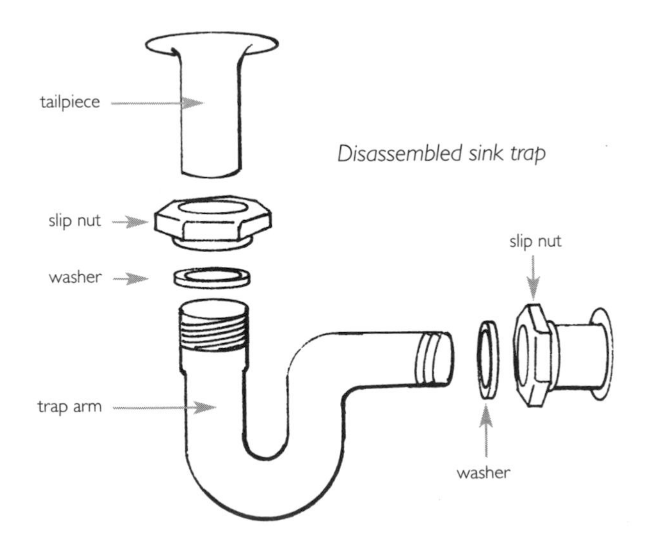

Disassembled sink trap

Fixing a Bathroom Pop-Up Drain Stopper

We have good news and bad news. The good news is that fixing a pop-up stopper is easy. The bad news is that you may have to remove everything that's accumulated under your sink. But you were dying to tackle that job anyway, weren't you?

A pop-up stopper can *stop* working if its parts wear out. The repair is easy, but the names of the parts are bizarre enough to be confusing, such as *clevis strap* and *pivot rod*. But don't fret, just think of how you'll blow your opponent out of the water the next time you play Scrabble.

To fix a pop-up stopper, it's important to first understand its parts. Look under the sink and, if necessary, shine a flashlight inside. With this book on the floor next to you, do a *search and find* for all of the parts of the

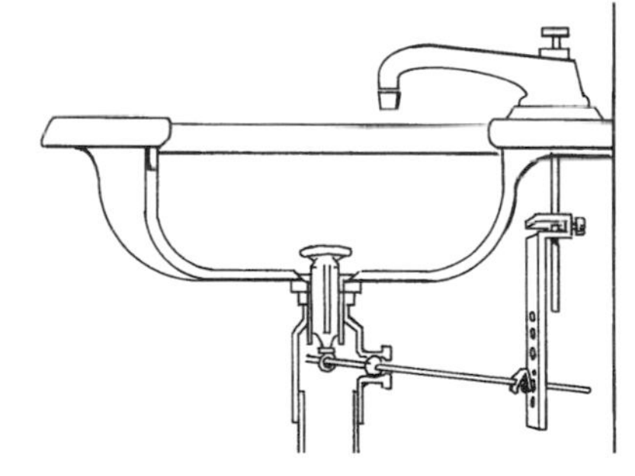

Pop-up stopper mechanism

Tools Needed

Lubricating spray

Slip-joint pliers 

Flashlight 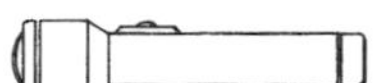

pop-up mechanism. Reach under the clevis strap and push it up to get a feel for how the pop-up stopper works.

Adjusting a Pivot Rod

The pivot rod is the narrow metal bar that fits into the clevis strap and is held in place by a spring clip. To adjust the pivot rod, simply pinch the spring clip and pull it off. Pull out the pivot rod and move it up or down into one of the holes in the clevis strap. Attach the spring clip to the pivot rod behind the clevis strap. If the pop-up still isn't sealing the drain hole quite right, then move the pivot rod again.

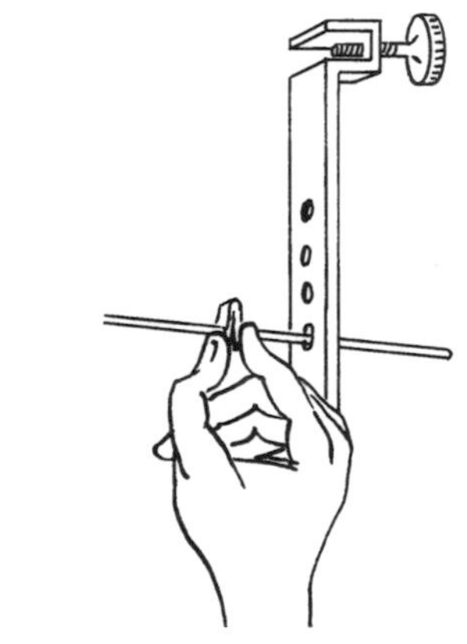

Pinching spring clip

Adjusting a Lift Rod

If the pop-up stopper still doesn't work, then adjust the lift rod. First, locate the clevis screw, which is typically hidden from view behind the clevis strand. You may need to refer to the illustrations while using the flashlight and feel around for it. If that fails, stick your head and shoulders inside the cabinet and look for it.

In some pop-up mechanisms, the clevis screw is a thumb screw, which you can easily loosen with your thumb and index finger by turning it counterclockwise. If the screw is stuck, apply the lubricating spray and use slip-joint pliers to loosen it. Push up the lift rod (attached to the clevis strap) to shorten it. Tighten the screw by turning it clockwise. Test the pop-up stopper and, if necessary, repeat the process.

Loosening a clevis screw that's not a thumb screw requires slip-joint pliers, a flashlight, and possibly lubricating spray. You may need to put your head and shoulders into the cabinet to get a good grip on the screw with the pliers. Turn the screw counterclockwise. If the screw won't budge, apply the lubricating spray before trying again.

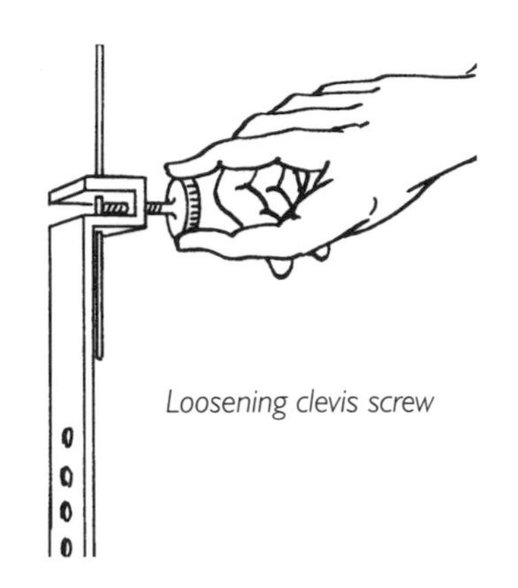

Loosening clevis screw

If you're still having difficulty, pinch the spring clip, remove the pivot rod, and turn the clevis strap so that it faces you rather than the wall. Hold the clevis strap with one hand and loosen the screw with the pliers.

Once the screw is loose, push up the lift rod (attached to the clevis strap) to shorten it. Tighten the screw with the pliers by turning it clockwise. Test the pop-up stopper and, if necessary, repeat the process.

Unclogging a Bathtub Drain

When Rebecca moved in with some male friends after college, it didn't take her long to realize that not only were the guys useless in the kitchen but they didn't know their screwdrivers from their wrenches. So, she unpacked her toolbox (a graduation gift) and went to work. She fixed the knob on her bedroom door and then tightened the handle on the toilet. Her roommates really stood up and saluted when she unclogged the bathtub drain. A new pecking order was immediately established among the roosters—the chick was in charge.

Have you devised a detailed timetable for showers in your house because your bathtub is perpetually clogged from soap and hair? Throw the chart away, because this repair is quick and dirty.

Tools Needed

Needlenose pliers  Flathead screwdriver 

Rag  Petroleum jelly 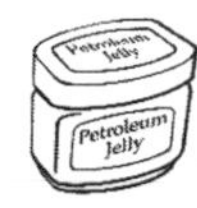

Plunger 

Tub with a Pop-Up Stopper

Flip up the drain lever, which is located on the overflow plate cover. Turn the stopper counterclockwise to loosen. Pull out the stopper and remove any accumulation of hair. If the water recedes down the drain, then you've solved the problem. Insert the stopper into the drain.

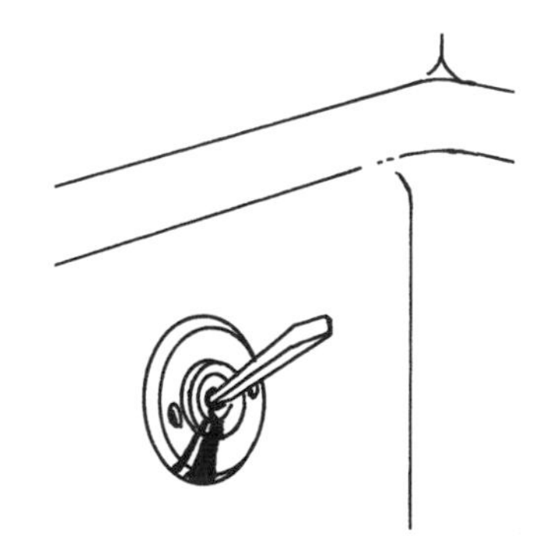

Drain lever in "up" position

Tub with a Strainer

Step into the tub and insert the needlenose pliers into two of the holes in the strainer. With a tight grip on the pliers, turn the strainer counterclockwise to remove it (you may need to use one hand on each handle of the pliers to exert more force). Once the strainer is out, remove any accumulation of hair. If the water recedes, then the problem is solved. Insert the strainer into the drain hole and turn it clockwise, making certain to get a tight seal.

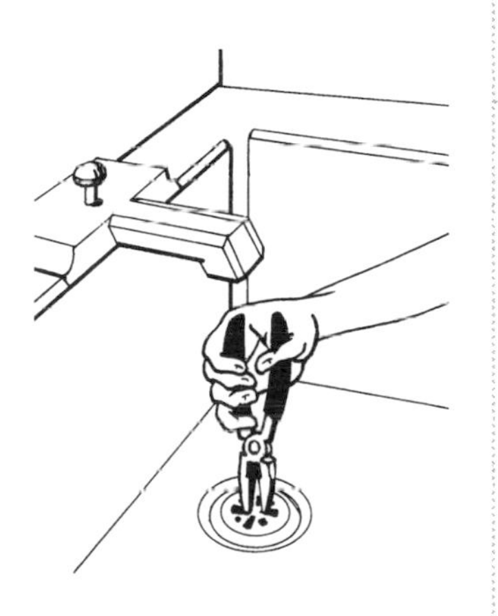

Removing strainer

If your tub has a non-removable strainer, you'll need to proceed to the repair on page 44.

If the Drain Is Still Clogged

Use the flathead screwdriver to remove the screw(s) from the overflow plate cover. Take off the cover and stuff a rag into the hole to provide greater suction.

Apply petroleum jelly to the rim of the plunger cup. Place the plunger directly over the drain hole. If the water doesn't cover the plunger's cup, add water to the tub. Plunge rapidly about a dozen times. Remove the plunger, allowing the water to recede down the drain.

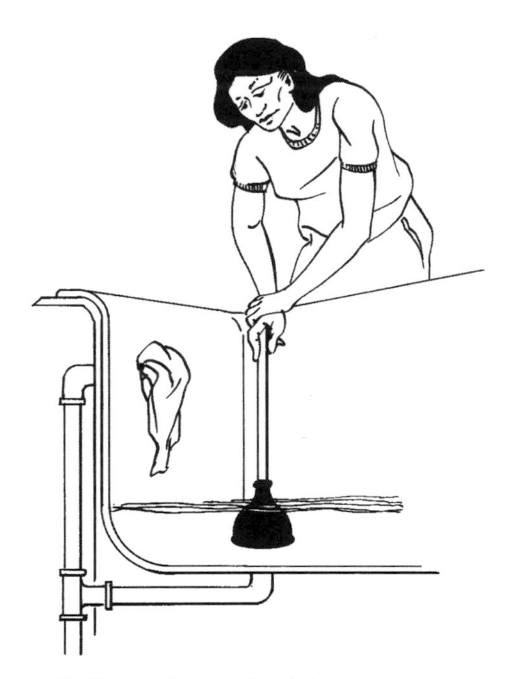

Stuffing rag into overflow hole

Replacing Sealant Around a Bathtub

Cristina and Alexandra have been neighbors for years, helping paint each other's houses, install dimmer switches, and clean gutters. So, when it was time for Cristina to recaulk her bathtubs, she asked her friend for some neighborly help. Together they removed and replaced the sealant around all three bathtubs, while sharing family stories, and when the job was done they shared the bill for a great lunch out.

Even if you're not having a friend help you with this project, you still need to clean the bathtub and tile before beginning. Sealants will adhere properly to the joint only if the area is free of mildew and dirt.

The place where the bathtub and tile meet is called the joint. This is where water from showers and baths can cause mildew and damage to tiles and walls, which could mean big bucks. Therefore, it's vital to protect the joint with a protective rubber-type sealant, such as silicone. You'll know when to replace the sealant when you see cracks or erosion.

Silicone is a permanent adhesive, which means that if it dries on your skin or clothing, it may be difficult to remove.

Tools Needed

Painter's tape 

Flathead screwdriver 

Tub and tile cleaner 

Rag 

Silicone with caulking gun 

Scissors 

Rubber gloves 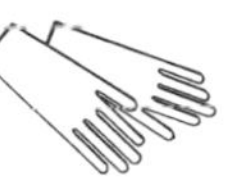

If you want to paint over the sealant, purchase a silicone whose label states it is paintable. As always, read the manufacturer's instructions before beginning.

Silicone comes in small tubes that you can squeeze like toothpaste to apply the caulk around the tub, or in larger tubes that fit into a caulking gun. If you have more than one bathtub in your home, stagger replacing the sealant on the others, because the drying time can be as much as 24 hours or more.

Removing Old Sealant

First, attach painter's tape above and below the joint to prevent scratching the tub and tile when you're removing the sealant. Use a flathead screwdriver to pry loose the sealant from around the tub. If the sealant is a silicone caulking, you'll be able to pull a lot of it out with your fingers; if it's grout, you'll need to scrape away at it with the screwdriver. Once all of the sealant has been removed, clean the area of any mildew or dirt and dry completely with a rag.

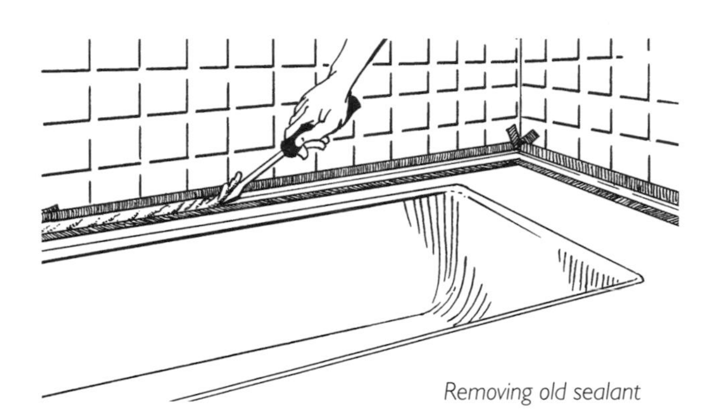

Removing old sealant

Preparing New Sealant

For sealant in a hand-held tube, remove the sealant's cap. Use the scissors to cut off the tip of the nozzle at an angle to produce the desired size of bead. For sealant in a large tube, create a hole in the opening by using the seal puncture on the caulking gun and load the tube into the gun.

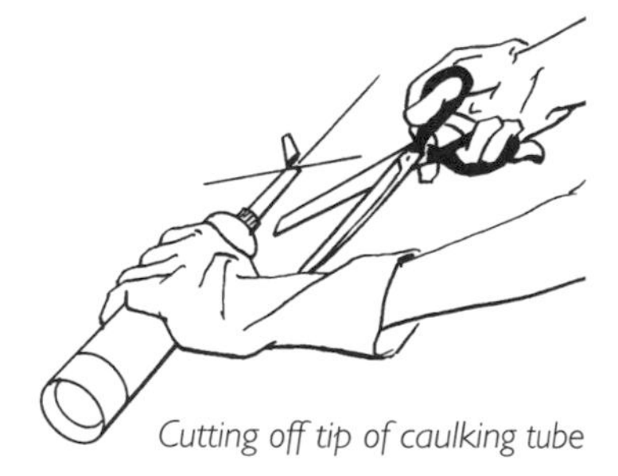

Cutting off tip of caulking tube

Applying New Sealant

Starting at one end of the tub, apply the silicone along the joint in a thin bead. Wearing rubber gloves, use one finger to gently smooth any uneven caulking. Remove the painter's tape. Dampen a rag and wipe away any caulking that went astray on the tiles or tub.

For recaulking around the base of the tub (where it meets the flooring), follow the same steps.

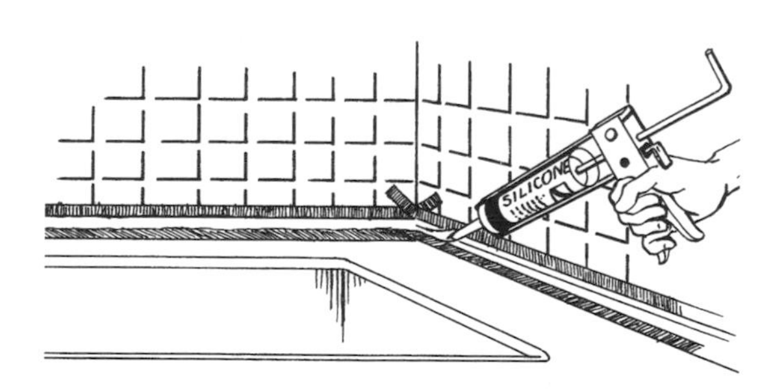

Applying caulking around tub

Faucets

Repairing a Leaking Faucet

The road to hell is paved with good intentions. So is the road to a plumber's wallet. Therefore, don't start this project without reading up on faucets first, and never tackle this project on a weekend or holiday when plumbers charge extra. Not that we don't have full confidence in you!

Because there are so many different styles of faucets, these repairs are for only the most common types.

You know the old saying "You can't tell a book by its cover"? Well, the same holds true for faucets. All faucets have a spout and one or two handles, but the differences among them are not based on looks but on what lies inside. Therefore, you might not know which type you have until you disassemble it. Of course, it's best to always take the parts to a plumbing supply or hardware store when purchasing replacements.

Faucets are divided into four categories based on their interior mechanisms: *compression, ball-type, cartridge*, and *disk*. A *compression* faucet, the oldest type, typically has two handles. Most kitchen faucets are either *ball-type* or *cartridge*. A *ball-type* faucet's handle rotates, while the handles of *cartridge* and *disk* faucets move up and down. A *disk* faucet, a combination of a ball-type and cartridge faucet, is the high end faucet type, because its disk assembly rarely has to be replaced.

Did you know that the location of the drip determines the cause, thereby determining the solution? We give repair instructions for faucets with leaks originating at the spout and for those with leaks from the handle/base.

Before doing any repair work on a faucet, remember to turn *off* the water at the sink's shut-off valve (located underneath the sink) or at the main shut-off valve before beginning. Turn *on* the sink and shower faucets to drain the water from the pipes on the level where you're working and the floor above, if applicable.

Compression Faucet

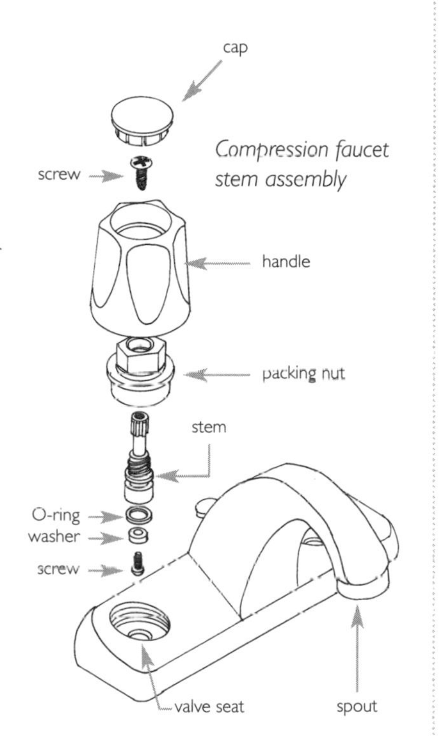

Compression faucet stem assembly

If the Faucet Leaks from the Spout

When the handles are turned *off* and water drips from the spout, then the problem may lie with a damaged washer, seat, or stem.

Turn *off* the water at the sink's shut-off valve (located underneath the sink) or at the main water supply shut-off valve. Cover the sink drain to prevent small parts from entering. Move the handle to the *on* position to drain any water.

Tools Needed

Metal fingernail file 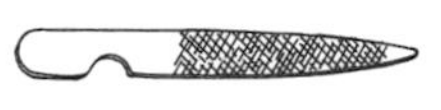

Screwdriver (Phillips or flathead) 

Faucet handle puller 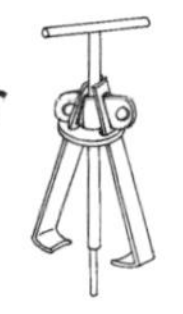

Slip-joint pliers 

Plastic sandwich baggie 

If your faucet has a cap on top, pop it off with the metal fingernail file. Use the screwdriver to remove the screw on each handle, and lift off the handles. If you are unable to remove a handle, use the faucet handle puller to extract it. Place the sidebars of the puller under the handle. Turn the bar at the top of the puller so that the shaft is inserted into the handle. Tighten the bar and pull out the handle.

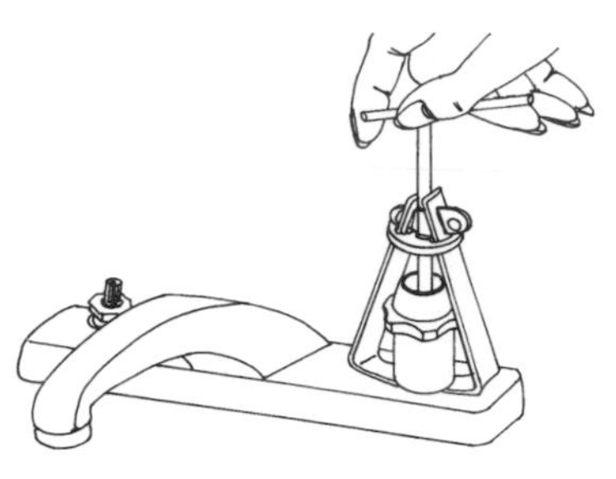

Handle puller inserted into faucet stem

To remove a stem, place the slip-joint pliers around the packing nut of the stem and loosen by turning it counterclockwise. Lift out the stem with your hand or pliers, and place inside a baggie. Repeat the procedure for the other stem. Take the baggie, with the two stems intact, to a plumbing supply or hardware store to purchase new stems, washers, or valve seats.

Installing a New Washer

Using the screwdriver, remove the stem screw and pry out the old washer. Replace it with the new washer. Insert the stem screw and tighten. Repeat the procedure on the other stem.

Removing stem screw

Tools Needed

Screwdriver (Phillips or flathead) 

New washer 

Installing a New Valve Seat

It's important to replace the valve seat while you're replacing a washer, because a worn valve seat will damage a new washer. To find out if the valve seat is worn, insert a finger into the faucet body (where the valve seat is located) and rub around the edge. If it feels rough, then the valve seat is deteriorated.

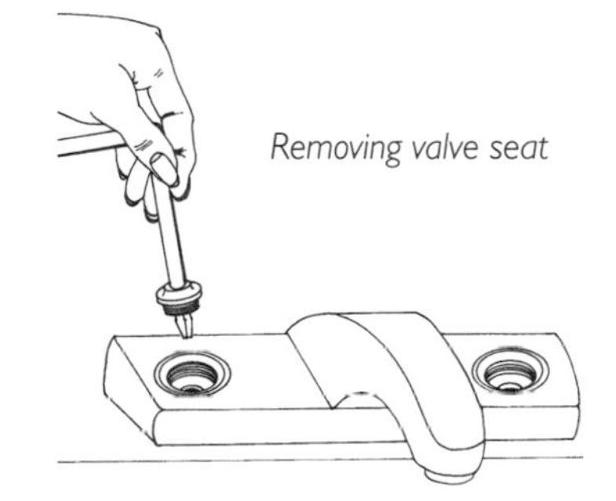

Removing valve seat

To remove the old valve seat, use the appropriate-size end of the seat wrench (L-shaped) and insert it into the faucet body. Turn it counterclockwise to loosen the valve seat. Remove it from the faucet body.

Apply several layers of Teflon tape counterclockwise onto the new valve seat. Place the new valve seat on the appropriate-size end of the seat wrench and insert it into the faucet body, avoiding an uneven threading. Turn it clockwise to tighten.

Teflon tape is a white thin pliable tape that wraps around pipe threads to prevent pressure leaks.

Tools Needed

Seat wrench

Teflon tape

New valve seat

Reinserting an Original Stem or Installing a New Stem with a New O-Ring

You have the option of replacing just the O-ring instead of the stem if you want to save money. However, since you have to take out the stem to remove the O-ring, it makes sense to install a new stem as well.

Before installing the new stem (or reinstalling the original), make sure it is in a fully retracted (i.e., open) position. If you are using the original stem and it came equipped with an O-ring, carefully cut it off, using a utility knife, and place the new O-ring on it.

Place the stem in the faucet body and tighten by hand. Use the slip-joint pliers to tighten the packing nut on the stem, turning it clockwise.

Before replacing the handles, turn the water *on* at either the sink's shut-off valve or the main water supply shut-off valve to test for leaks. If leaks occur, it could be that the packing nut simply needs another quarter turn. Or the problem could be that the stem was not properly installed in the faucet body.

Turn *off* the water again. Tighten the packing nut, making sure that you don't overtighten, because doing so could restrict movement of the stem and make it difficult to turn the faucets on and off. If so, then remove the packing nut and apply graphite packing string on the nut and replace.

If the stem is causing the problem, remove and reinsert it, being careful to evenly thread it into the faucet body. Tighten securely without overtightening.

Tools Needed

Utility knife

New O-ring 

New stem (includes a new O-ring) 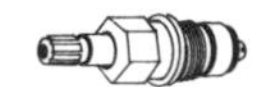

Slip-joint pliers 

Graphite packing string 

If a Faucet Leaks from the Handle/Base

We know this is weird, but the instructions (including tools) for this repair are the same as for "Reinserting an Original Stem or Installing a New Stem with a New O-Ring" (page 52). The reason is that the O-ring is located on the stem and the only way to replace it is to remove the stem. So, follow all of the above instructions to the very last *drop!*

Ball-Type Faucet

If the Faucet Leaks from the Spout

When the handle is turned off and the faucet leaks from the spout, the problem could be a loose adjusting ring, a damaged ball, worn valve seats, or worn springs. Before replacing any parts, first try tightening the adjusting ring.

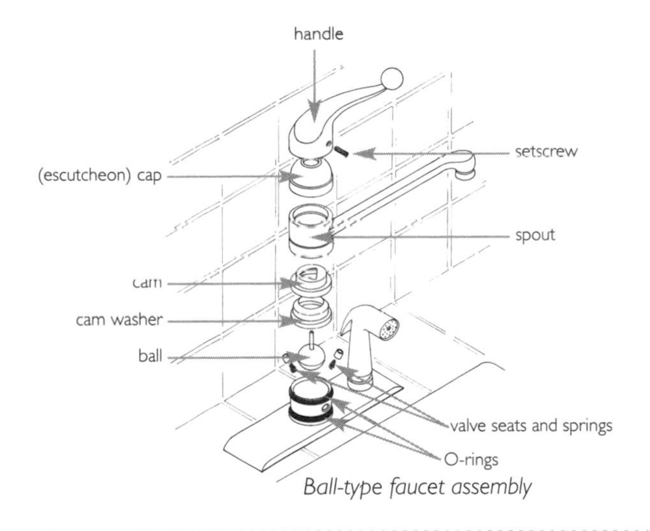

Ball-type faucet assembly

Tools Needed

Allen wrench set (a.k.a. folding hex key set)

Cam tool (found in most faucet repair kits)

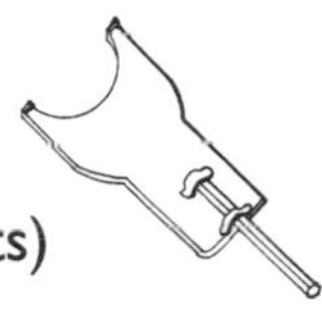

Tightening the Adjusting Ring or Dome Housing

Find the right-size Allen wrench to fit into the hole underneath the handle. Turn the Allen wrench counterclockwise to loosen the setscrew. Lift off the handle.

Insert the cam tool into the adjusting ring (located in the cap) and turn clockwise to tighten, making certain not to overtighten and crack the handle.

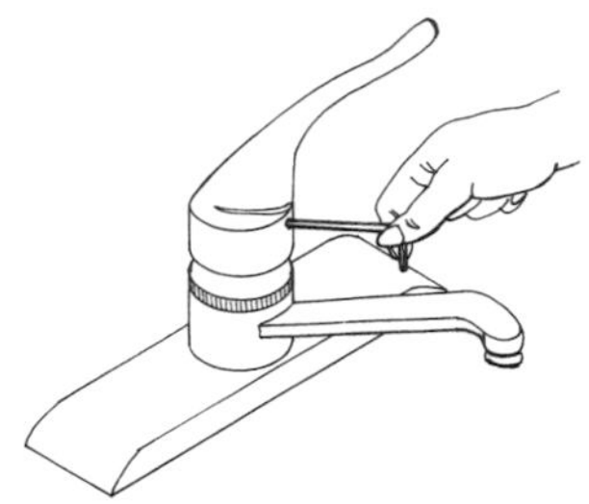

Removing setscrew

Replace the handle and setscrew and test for leaks. If the spout still leaks, then you'll need to replace the worn parts.

Not all ball-type faucets have an adjusting ring or dome housing.

Replacing a Damaged Ball, Valve Seats, and Springs

Turn *off* the water at the sink's shut-off valve (located underneath the sink), or at the main water supply shut-off valve. Cover the sink drain to prevent small parts from entering. Move the handle to the *on* position to drain any water.

Find the right-size Allen wrench to fit into the hole underneath the handle. Turn the Allen wrench counterclockwise to loosen the setscrew. Lift off the handle. Tape the screw to the handle, so you won't lose it.

Tape the head of the slip-joint pliers to prevent marring. Place the pliers on the grooves of the cap, turning counterclockwise to loosen. Lift off the cap.

Pull out the cam, cam washer, and ball with your fingers (noting the position of the ball) and place in a baggie. If the ball is worn, you'll need to replace it (purchase only a metal ball, because it will last longer).

Using a screwdriver, remove the springs and valve seats from the faucet body and place in a separate baggie. Take the springs and valve seats (and the worn ball, if applicable) to a plumbing supply or hardware store and purchase replacements.

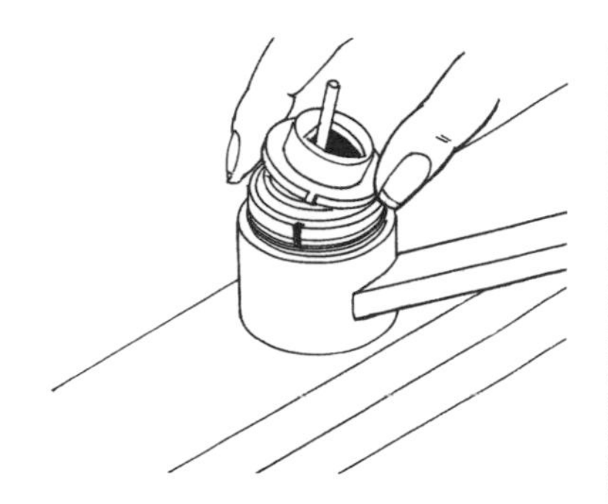
Aligning notches on cam and faucet body

Insert the new springs and valve seats into the faucet body. Position the ball the same way you found it. Place the cam washer and cam back in the faucet, making sure that the notch on the cam fits into the slot on the faucet body. Reattach the cap and tighten with the adjustable wrench. Replace the handle and insert the screw. Tighten it with the Allen wrench. Turn the water *on* at the sink's shut-off valve or the main water supply shut-off valve.

If the Faucet Leaks from the Handle/Base

Damaged O-rings are to blame and will need to be replaced.

Tools Needed

Allen wrench 

Masking tape 

Slip-joint pliers 

Plastic sandwich baggie 

Screwdriver (flathead or Phillips) 

Replacing the O-rings

Turn *off* the water at the sink's shut-off valve (located underneath the sink), or at the main water supply shut-off valve. Cover the sink drain to prevent small parts from entering. Move the handle to the *on* position to drain any water.

Find the right-size Allen wrench to fit into the hole underneath the handle. Turn the Allen wrench counterclockwise to loosen the setscrew. Lift off the handle.

Tape the head of the slip-joint pliers to prevent marring. Place the pliers on the grooves of the cap, turning it counterclockwise to loosen. Lift off the cap.

Pull out the cam, cam washer, ball, valve seats, and springs with your fingers (noting the position of the ball) and place in a baggie.

Remove the spout by twisting it upward. Use the utility knife to cut off the O-rings. Install the new O-rings, replace the spout, and reassemble the faucet. Turn *on* the water at the sink's or main water supply shut-off valve and test for leaks.

Tools Needed

Allen wrench

Masking tape

Slip-joint pliers

Plastic sandwich baggie

Utility knife

New O-rings

Cartridge Faucet

If the faucet leaks from the spout, the problem lies with the cartridge. If the leak originates at the handle/base, then worn O-rings are to blame. Typically this is where we'd tell you to replace either the worn cartridge or the O-rings, but because the O-rings are located on the cartridge, it makes sense to replace both, no matter the source of the leak.

Turn *off* the water at the sink's shut-off valve (located underneath the sink) or at the main water supply shut-off valve. Cover the sink drain to prevent small parts from entering. Move the handle to the *on* position to drain any excess water.

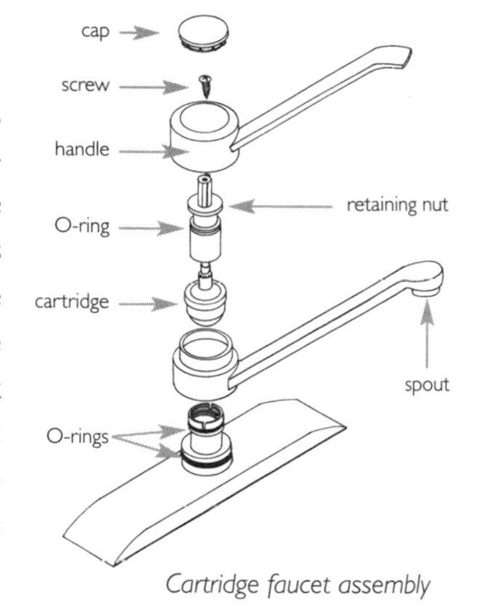

Cartridge faucet assembly

Tools Needed

Metal fingernail file 

Screwdriver (flathead or Phillips) 

Masking tape 

Slip-joint pliers 

New O-rings 

Plastic sandwich baggie 

New cartridge

Removing the Old Cartridge

Pop off the cap with the metal fingernail file. Use the screwdriver to loosen the screw. Lift up and tilt the handle to remove it. Pry off the retaining nut with the screwdriver or, if it's stuck, use slip-joint pliers (with the head wrapped in masking tape to prevent scratching the surface). Place the pliers on the retaining nut (ring), turning it counterclockwise to loosen. If there is a retaining clip (U-shaped) located on the side of the stem, remove it with pliers. Tape the retaining clip to the handle for safekeeping.

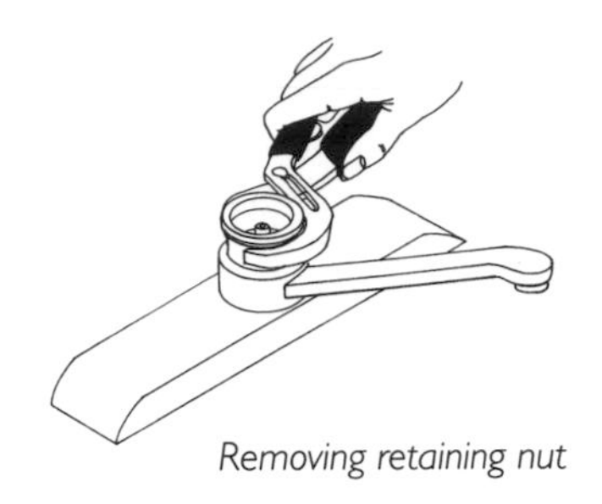

Removing retaining nut

Grasp the top of the cartridge with the slip-joint pliers and pull the cartridge straight up and out. Place the cartridge inside a baggie and take it to the plumbing supply or hardware store for replacement.

Removing the Original O-rings

To remove the spout, twist upward and pull it off. Using the utility knife, carefully cut the worn O-rings from the faucet body.

Installing New O-rings and Cartridge

Place the new O-rings on the grooved parts of the faucet body. Install the spout and then the cartridge. It's really important that you place the new cartridge so that its tab faces forward.

If there is a retaining clip, replace it now. Return the retainer nut and use the slip-joint pliers to tighten it by turning it clockwise. Replace the handle, screw, and cap. Turn *on* the water and check for leaks. If the hot and cold water are reversed (*oops!*), go back and turn the cartridge 180 degrees.

Disk Faucet

A disk faucet usually leaks because of deposit buildup on the rubber seals or between the ceramic disks. If the faucet still leaks after you've cleaned the seals, you'll need to replace the entire disk assembly.

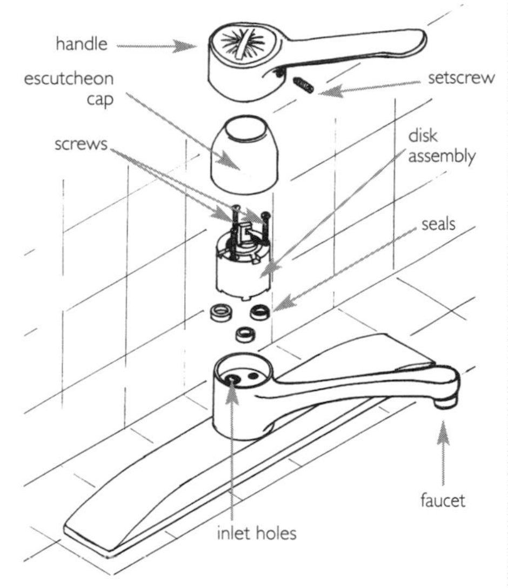

Disk faucet assembly

Cleaning the Seals

Turn *off* the water at the sink's shut-off valve (located underneath the sink), or at the main water supply shut-off valve. Cover the sink drain to prevent small parts from entering. Move the handle to the *on* position to drain any excess water.

Tools Needed

Allen wrench set 

Masking tape 

White vinegar 

Slip-joint pliers 

Plastic sandwich baggie 

Screwdrivers (flathead and Phillips) 

Scouring pad (non-metallic) 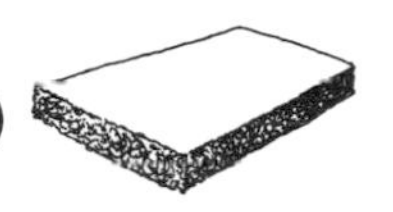

Find the right-size Allen wrench to fit the setscrew in the handle. Turn the Allen wrench counterclockwise to loosen the screw. Lift off the handle. Tape the head of the slip-joint pliers to prevent scratching the finish. Place the pliers on the escutcheon cap (dome housing), turning it counterclockwise to loosen. Remove the cap.

Use the Phillips screwdriver to remove the 3 screws located in the disk assembly, and pull the assembly out while noting its position. Carefully remove the 3 rubber seals on the bottom of the cylinder, using the flathead screwdriver. If the seals aren't worn, clean them with a non-metallic scouring pad or toothbrush. Soak the cylinder in vinegar to eliminate deposit buildup in the inlet holes, which house the seals.

Removing rubber seals

If the seals are worn, place them with the cylinder in a baggie and take them to a plumbing supply or hardware store for replacement.

Installing the Seals

Insert the original or new seals into the inlet holes. Place the disk assembly in the faucet body, aligning the holes on the bottom of the disk assembly with the holes in the faucet body. Return the 3 screws and tighten them with the Phillips screwdriver. Replace the escutcheon cap and handle, and fasten the setscrew into the handle with the Allen wrench.

After you've returned all of the faucet parts, it's extremely important to turn the handle to the *on* position ***before*** turning the water supply *on*, because a sudden surge of air can crack the ceramic disks. So, the best thing to do is to turn *on* the water supply *slowly*.

If the faucet still leaks, the disk assembly needs to be replaced. Take the old disk assembly to the plumbing supply or hardware store for replacement. (Note: The new disk assembly will include seals.)

Unclogging a Sink Spray

Grace got tired of writing honey-do lists only never to see anything scratched off. So, on the first Sunday of the football season, she left her husband in front of the TV and went to the local hardware store. She returned with a game plan and equipment to fix the sink spray that hadn't worked since the Miami Dolphins had a perfect record. During the football season, she tackled a project every week. Her husband, thrilled by her new talent, called her the "Repair Fairy," because things "magically" got fixed around the house. Grace smiled and said, "That's nice, honey. Just don't think about leaving your teeth under the pillow."

A sink spray is a great kitchen gadget . . . when it works. When it doesn't, you can score a touchdown, just like Grace, by doing this easy repair.

Tools Needed

Small flathead screwdriver 

Plastic sandwich baggies 

White vinegar 

Small bowl 

Old toothbrush 

Toothpick 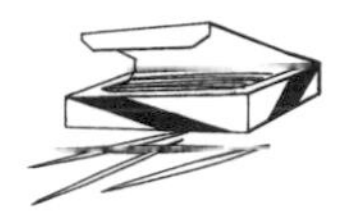

The difference between an older model sink spray and a new one is simply in how you clean it. To clean a new sink spray, you have to disconnect it at the hose; to clean an older model, you have to dismantle the head. If in doubt about which kind you own, try to disassemble the sink spray to find out.

Unclogging an Older Model

A sink spray will have either a screw that's visible, or one that's hidden behind a cover. If the screw is hidden, use the small screwdriver to pry the cover off, and remove the screw. Take out the perforated disk, sleeve, and washers and place all of these parts along with the screw in a baggie. (You may feel more comfortable putting each part in a separate baggie and labeling each baggie with a number corresponding to the part you removed first.)

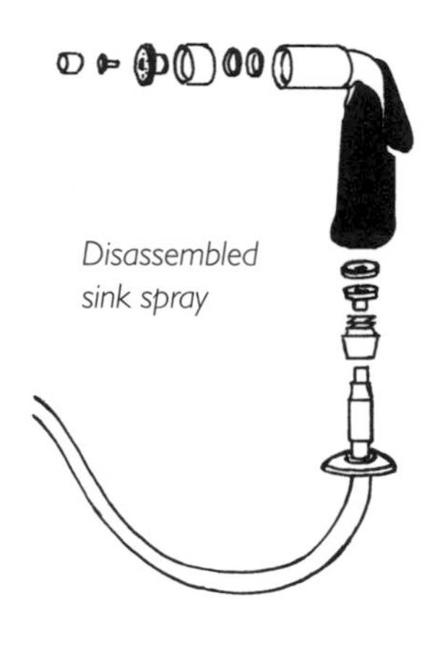

Disassembled sink spray

Soak the cover in vinegar for at least 1 to 1½ hours. Remove the cover and scrub gently with a toothbrush. Poke the holes with a toothpick and rinse with warm water. Replace the parts in the opposite order, starting with the washer.

Unclogging a Newer Model

Turn the spray head counterclockwise to unscrew it from the hose. Soak the entire spray head in a bowl of vinegar for at least 1 to 1½ hours. Poke the holes with a toothpick. Rinse the spray head with warm water and screw it back onto the hose.

Spray head soaking in vinegar

Replacing a Showerhead and Shower Arm

While lying in your bed at night, are you counting water droplets instead of sheep? Stop the water torture and replace your broken showerhead.

Replacing a showerhead, whether because it's leaking or because you want a fancier model, is an easy project. But, just as with any plumbing repair, stuff happens—a nut may be hard to loosen, a leak suddenly appears, etc. Don't panic, because even the most seasoned plumber can encounter unexpected problems.

We recommend you remove the showerhead and take it to the hardware store before purchasing another—not because we want you to be stuck with the same style, but because a new showerhead needs to fit onto the existing shower arm. Or, you can replace the shower arm to fit the new showerhead. It's easy to do . . . really!

Tools Needed

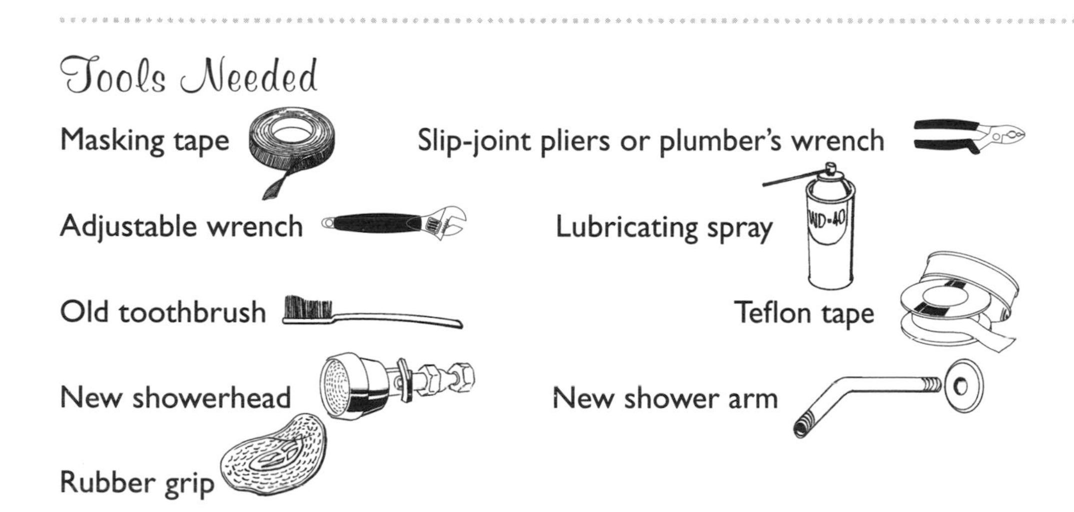

Teflon tape is white thin pliable tape that wraps around pipe threads to prevent pressure leaks.

Replacing a Showerhead

Removing an Old Showerhead

Before starting, wrap masking tape around the heads of the slip-joint pliers and adjustable wrench to prevent marring. Attach the pliers to the top of the shower arm to keep it from rotating. Position the adjustable wrench on the collar nut and turn it counterclockwise. If the nut is difficult to unscrew, apply a lubricating spray to it. Once the nut is loose, you can continue unscrewing it with your hand. Clean off any residue on the threads with a toothbrush.

Slip-joint pliers and adjustable wrench on shower arm

Installing a New Showerhead

Apply the Teflon tape to the exterior threads of the arm, wrapping it counterclockwise until you have 3 or 4 layers. Before installing the new showerhead, make sure that the washer is located inside it. Put the new showerhead on the end of the shower arm, turning it clockwise with your hand. Place the slip-joint pliers on the middle of the arm and secure. Put the rubber grip on the collar nut and position the adjustable wrench on it. Turn the nut clockwise to tighten.

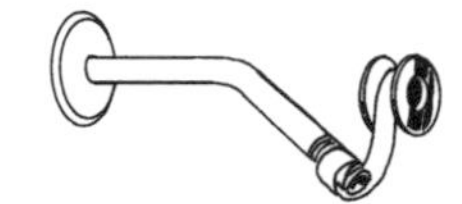

Applying Teflon tape to threads of shower arm

Securing new showerhead

Turn *on* the shower and watch for leaks at the collar nut. If there aren't any, *bravo*! If there are, don't fret. Just try tightening the collar nut again.

Replacing a Shower Arm

Removing an Old Shower Arm

Place both hands on the shower arm and turn it counterclockwise. If this doesn't loosen it from the wall, apply a lubricating spray to the base of the arm (where it connects to the wall). Try twisting it again, or use an adjustable wrench.

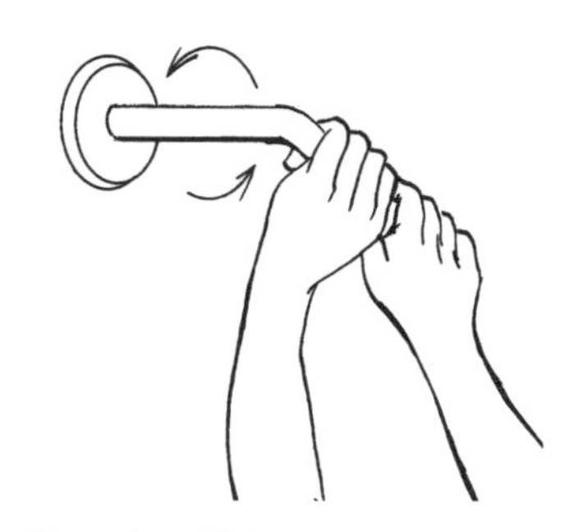

Removing old shower arm

Installing a New Shower Arm

Before starting, wrap the head of the adjustable wrench with masking tape to prevent any marring of the new pipe. Next, apply 3 or 4 layers of Teflon tape counterclockwise around the exterior threads on both ends of the arm. Insert the arm into the connection in the wall and twist clockwise. Use the adjustable wrench to tighten the shower arm.

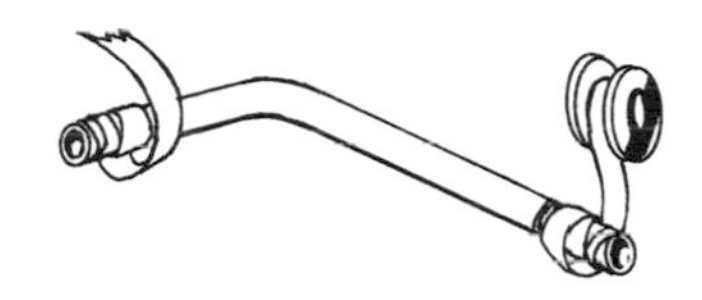

Applying Teflon tape to threads of shower arm

Insert the new showerhead onto the arm, twisting it clockwise. Use the adjustable wrench to tighten the head at the collar nut.

Turn *on* the shower and look for leaks at the 2 connections. If there are leaks, retighten the arm or head with the wrench.

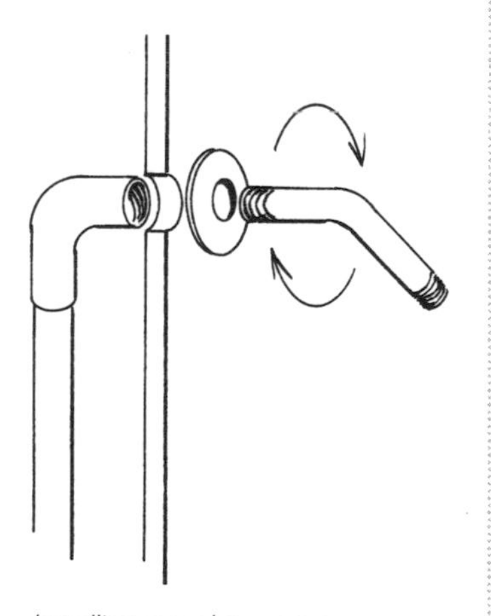

Installing new shower arm

If you need to replace a shower-head along with a shower arm, you might as well not separate them. Just leave the shower assembly in one piece, remove, and take to a plumbing supply or hardware store for replacement.

Seasonal Repairs

Winterizing Plumbing

As physicians, Alice and her husband fixed ailing people. But if something needed fixing around their house, they'd call a contractor and flip a coin to see who'd have to stay home. Alice realized they were spending too many of their precious vacation days with the repairperson instead of with each other. So, before the harsh winter arrived, she decided to do some preventive maintenance by first draining the water from her home's exterior faucets and hoses. Now, instead of thawing frozen pipes this cold winter, she and her husband will be defrosting on a warm, sunny beach.

You may not be ready to pull out your winter clothing, but don't put off covering those exposed water pipes in your home and winterizing your exterior water faucets.

Tools Needed

Lubricating spray

Adjustable wrench 

Rag 

Cold weather faucet cover 

Preparing Exterior Faucets

Whether you live in a new home with a frost-free exterior faucet or in an older home with a good ol' hose bib (common exterior faucet), your faucets are vulnerable during a harsh winter. The most important thing you can do to prevent faucets and pipes from freezing is to rid them of any water *before* winter.

First, you'll need to know if your home has an interior shut-off valve for the exterior faucet. Typically, the valve is located on the interior wall of the house, directly behind the exterior faucet, and will have either a gate valve (wheel-shaped) or a ball valve (lever). To shut it *off*, turn the gate valve clockwise or move the ball valve a quarter turn. If the gate valve is difficult to move, apply lubricating spray and use the adjustable wrench.

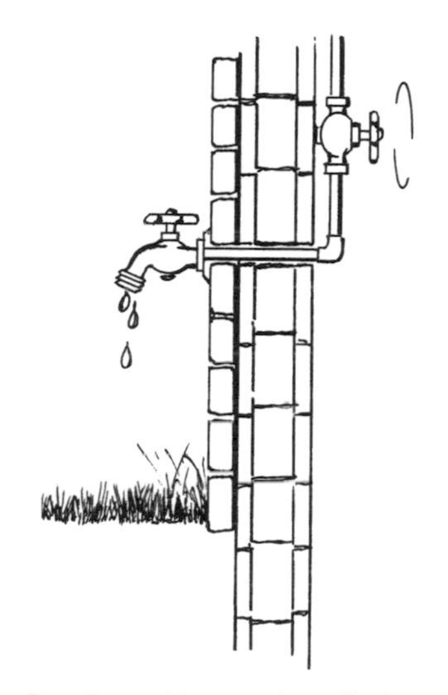

Exterior and interior shut-off valves

Go outside to the exterior faucet. Detach the garden hose, if applicable. Open the exterior faucet by turning the handle counterclockwise, allowing any water remaining in the pipe to drain.

Go back inside and open the bleeder cap (3/8"-round brass cap) located on the side of the main water supply valve. Using the adjustable wrench, slightly turn the bleeder cap counter-

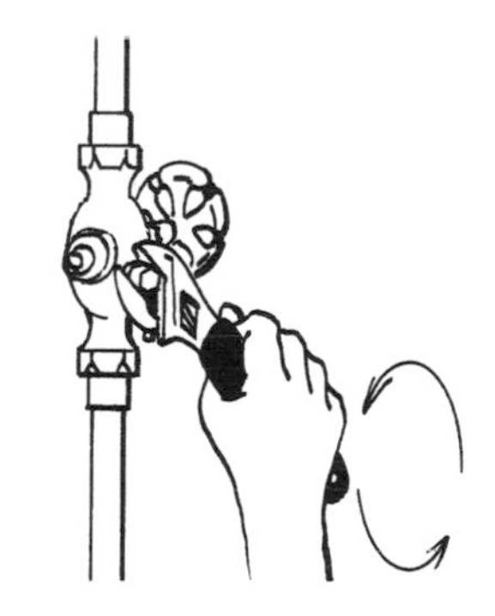

Opening bleeder cap on main water supply valve

Even if you have a frost-free exterior faucet, you still need to properly maintain it for the cold weather, because it's not 100 percent fullproof.

clockwise. Leave it open for several seconds to allow air into the pipes to help force any trapped water out. Use a rag to catch drops of water that may drain from the bleeder. Close the cap immediately.

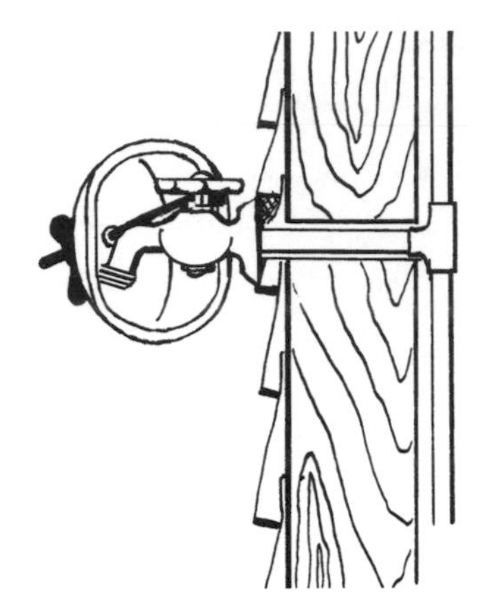
Installing cold weather faucet cover

After the water has drained from the exterior faucet, shut it *off,* turning it clockwise, to prevent cold air from entering the water pipes. For added protection, place the cold weather cover over the faucet and secure the hook (inside the cover) around it. Turn the wing nut, located on the top of the cover, until it's tight.

Drain any water that may still be inside the garden hose by hanging it over your deck or fence. If you have neither, extend the hose its entire length. Raise one end to waist level and putting one hand in front of the other, walk the length of the hose, draining it of any trapped water. You may have to repeat this exercise a few times.

If you don't have an interior shut-off valve, have one installed by a licensed plumber.

Insulating Interior Water Pipes

Installing insulation on exposed interior pipes, typically found in an unfinished basement, helps to prevent the pipes from freezing. There's

Tools Needed

Interior pipe insulation (tube)

Utility knife 

Duct tape 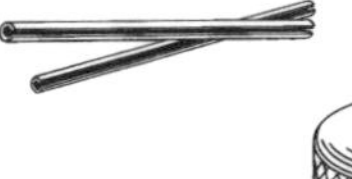

no need to remove the insulation after winter, because it also aids in reducing the energy used by the water heater to warm the cold water, as well as in muffling the sounds of noisy pipes.

Interior pipe insulation comes in different materials, lengths, and diameters, so don't go to the hardware store without knowing the size of your pipes. The standard size of residential pipes is ½ inch and ¾ inch. If you don't know the size of the pipes, use a tape measure to find the correct diameter.

Installing Insulation

The insulation tube is pre-cut down the center so that you can easily place it around the pipe. Peel off the tape found on both edges and stick the two together, working your way down the length of the tube. If you will be using more than one insulation tube, you may need to cut the tubing to size with the utility knife. Cut the duct tape and wrap the pieces where the tubes meet.

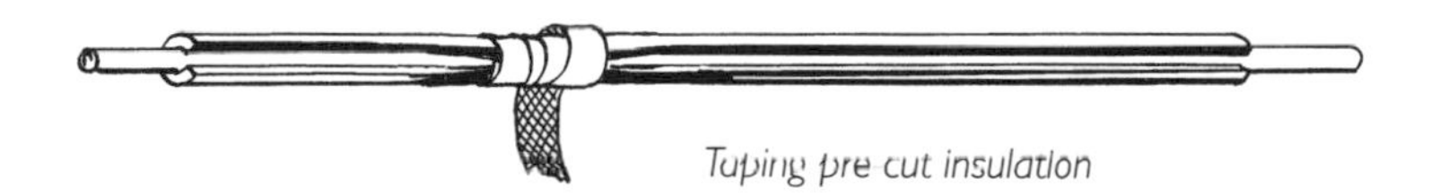

Taping pre-cut insulation

Thawing a Frozen Water Pipe

Some women go to Jamaica to get their groove back—Yvonne was going there to get one. While she was chillin' in the sun, her home's water pipes were chillin' in a record cold spell. The pipes burst, spewing tons of water into her home. The water rose above the furnace and went out into the street, where it iced over, forcing the city to close the road. Yvonne returned to a condemned house, angry neighbors, and a water bill equal to her Jamaican trip.

If your home is over 20 years old *and* you live in an area with harsh winters, you've probably had experience with frozen water pipes (hopefully not as bad as Yvonne's).

Water in a pipe can freeze when two things occur simultaneously: 1) the outside temperature goes below freezing; and 2) an exterior or interior water pipe is not properly insulated.

There will be no mistaking when a water pipe has burst in your home, but how can you tell when a pipe is freezing or has frozen? A pipe that is freezing emits *extremely* cold water. If a pipe is frozen, then *no* water will come out of the faucet.

Tools Needed

Flashlight

Hair dryer

Extension cord

If a Pipe Has Burst

An emergency fix for a burst water pipe is to immediately shut *off* the main water supply valve (see "The Main Water Supply Valve: How to Find It and Shut It Off," page 6). If there is standing water, do *not* turn on any electrical switches—use a flashlight instead. Contact a certified plumber... and your insurance agent!

If a Pipe Is Frozen

Use a hair dryer (with an extension cord, if necessary) to heat a frozen pipe. Because it's almost impossible to see where the freeze originated, you'll have to heat the entire length of the pipe. This could take as long as 20 to 30 minutes. Turn the faucet *on* and wait about 3 minutes. If water is still not coming out, try reheating the pipe.

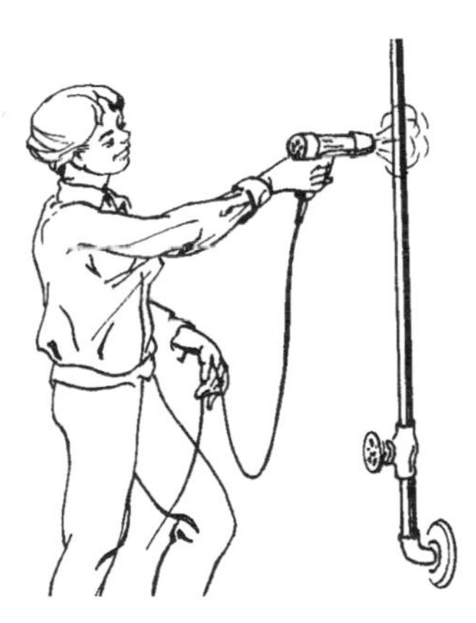

Heating frozen pipe with hair dryer

The best preventive measures in freezing weather are to leave a faucet dripping at a slow rate, open the cabinets to allow the warm air in the room to reach the pipes exposed to the exterior wall, and wrap the exposed interior water pipes with insulation (see "Winterizing Plumbing," page 66).

Clearing a Gutter and Downspout

We recommend you hire a contractor to do repairs such as fixing rust spots, holes, and leaking seams, tightening the fasteners that secure the gutter system to the house, and adjusting the slope of the gutter.

The first home repair Olivia ever did was out of necessity, because as a single parent raising four children on limited funds, a leaky roof didn't fit into her budget. She asked her neighbors for help, but none of the women had ever been on a ladder, let alone a roof. As Olivia was climbing up, her neighbors were yelling for her to stop her nonsense and get down. But when she held up the cause of the leak—an overthrown newspaper which clogged the gutter—they shouted, "You go, girl!' Olivia's 15 minutes of fame came and went (as well as the leak), but her gutters are still going strong.

Gutters and downspouts are the main parts of the exterior drainage system of your house. This system can carry thousands of gallons of water every year, so it's important to properly maintain it—once in the spring and again in the fall. Don't be fooled into thinking a heavy rain will clean the gutters for you, because it can actually make things worse. Decayed leaves and mud can block the flow of water in the gutters and downspouts and corrode them with mineral deposits.

Cleaning the gutter system is not a difficult job; however, it can be time consuming, depending on the number of gutters and downspouts you have and on the amount of accumulated debris. If the gutter is clogged, you'll have to clean it in sections, which means moving the ladder and climbing up and down it several times.

For safety reasons, don't tackle this job if the ground is wet or if it's very windy outside, be sure to wear shoes with good traction, and follow the rules for ladder safety (see "Practicing Ladder Safety," page 229).

Clearing a Gutter

Decide which section of the house you're going to tackle first. With your friend's help, place the ladder at the end of the gutter farthest from the downspout. Using a tape measure, check that the ladder is correctly angled against the exterior wall (see "Practicing Ladder Safety," page 229).

Climb the ladder, carrying the garden hose, bucket, and trowel. Go high enough so that you can see into the gutter and hook the nozzle of the hose to the rung above you. If your gutter has a leaf guard, flip it up, or remove and place on the roof above where it was located.

If you find a small amount of debris in the gutter, spray it toward the downspout. If the gutter is clear, secure the leaf guard into place.

Removing leaf guard

For a large amount of debris, remove it with

Shoes with good traction 

Helpful friend

Ladder 

Tape measure 

Garden hose with spray nozzle

Bucket or a plastic grocery bag 

Garden trowel or scoop

the trowel and place the debris in the bucket. Climb down the ladder and empty the debris-filled bucket into a trash can. With your friend's help, move the ladder to the next section of gutter to be cleaned. Work your way down the entire length of the gutter until it is completely cleared of debris. Secure the leaf guard into place.

Placing gutter debris into bucket

With your friend's help again, move the ladder back to the end of the gutter farthest from the downspout (the same spot where you started). Climb the ladder with the garden hose and spray any remaining debris toward the downspout. If there is standing water in the gutter, then the downspout may be clogged or the gutter is not properly sloped.

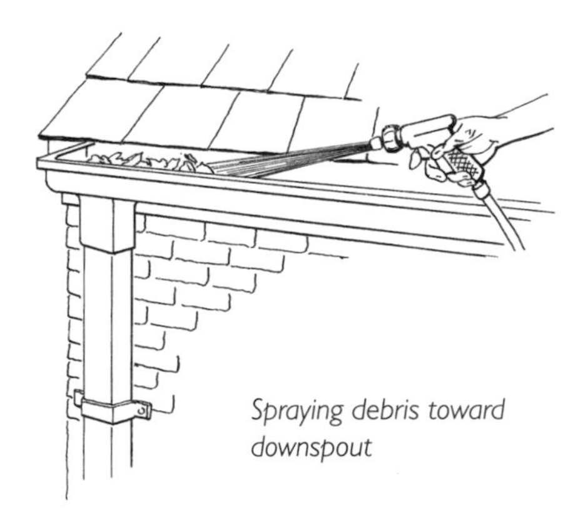

Spraying debris toward downspout

Repeat this procedure for all of your home's gutters.

Clearing a Downspout

With your friend's help, move the ladder close to the downspout. Climb the ladder, carrying the garden hose, going high enough so that you can feed it into the downspout.

Have your friend turn *on* the water, and wash the debris down the spout. If there is still standing water in the gutter, then the gutter may not be sloped enough to allow for proper drainage. We recommend you hire a contractor to adjust the gutter's slope.

Walk around to every downspout and make sure that each one has a splash block and that it's positioned correctly so that water is directed away from the house, not toward it.

Flushing debris through downspout

Downspout with splash block

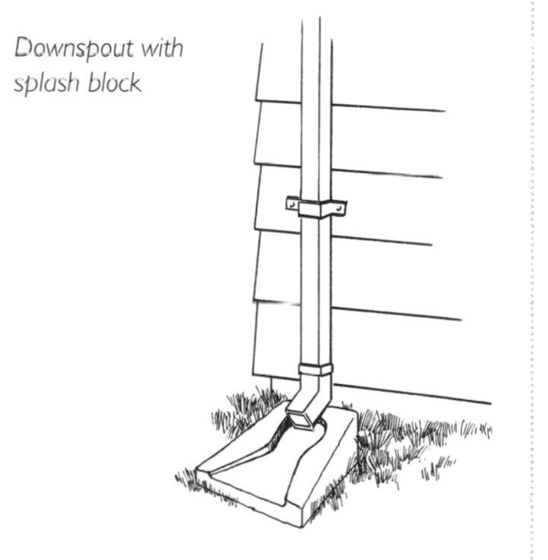

Remember to wear shoes with good traction and follow the rules for ladder safety (see "Practicing Ladder Safety," page 229).

Tools Needed

Shoes with good traction

Helpful friend

Extension ladder

Garden hose (without spray nozzle)

Bleeding a Hot Water Radiator

If your radiator is whistlin' while it's workin', that's not because it's happy. Whistling is a sign that it needs maintenance.

Always keep furniture away from a radiator and never paint it, as paint will block the flow of air.

A water system heats the home by bringing hot water from the boiler through the pipes and into the radiators or convectors. You can see by the illustrations that radiators and convectors differ in appearance, even though they provide the same service.

To help this heating system run efficiently, you need to clean it regularly and bleed it every fall. To bleed a radiator means to remove the hot air trapped inside that is blocking the flow of heat. Older radi-

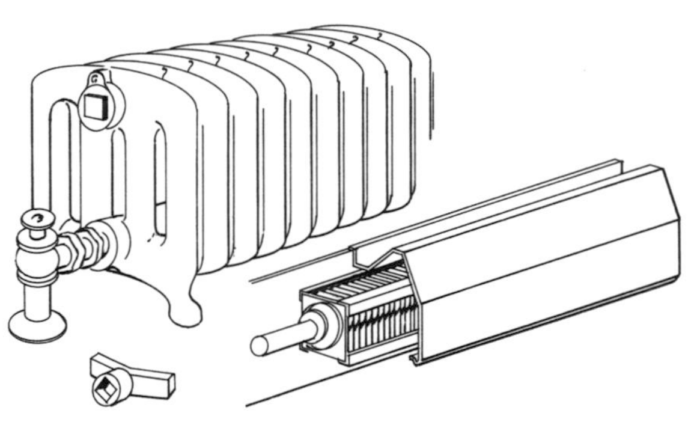

Radiator and convector

Tools Needed

Rag 

Vacuum with brush attachment 

Needlenose pliers 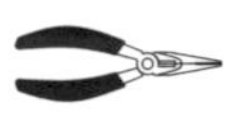

Cup 

Valve key or flathead screwdriver 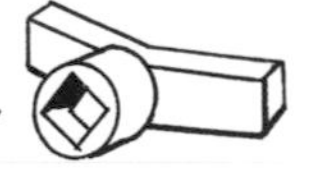

ators typically require a valve key to turn the bleeder valve; if not, a screwdriver may work.

If your hot water system uses convectors, check for any bent fins and straighten them with needlenose pliers. Some convectors have the same bleed valves as radiators; if yours doesn't, contact a certified plumber to have the system professionally maintained.

Use the rag and vacuum with brush attachment to clean all of the radiators before bleeding them.

Turn the thermostat up high enough so that the furnace turns *on*. If you live in a multistoried house, start with the radiator on the highest level; if you don't, start with the one farthest from the boiler. With one hand holding the cup underneath the bleeder valve, insert the valve key or screwdriver into the valve and turn it counterclockwise to open.

Once the hissing sound of escaping air stops, water will begin to dribble out. Quickly turn the valve *off*, being careful to avoid the hot steam and water that may spew out. Repeat this process on all of the radiators or convectors.

During the winter months when the heat is on in your home, periodically check the radiators to see if any are cold to the touch. Bleed any that are, starting with the one farthest from the boiler. If a radiator still is not producing heat, contact a certified plumber.

Bleeding radiator

If you live in an apartment, remember that landlords are required by law to provide heat—not air conditioning, though. If the heat in your apartment isn't working, first notify your landlord. If this doesn't bring about a change, contact your city government.

Electricity

Circuit Breakers vs. Fuses

- Understanding Circuit Breakers and Fuses
- Creating a Circuit Map
- Restoring Electrical Power
- Testing a Receptacle Ground Fault Circuit Interrupter (GFCI)

Light Bulbs

- Removing a Broken Bulb from a Light Fixture
- Replacing a Light Bulb in a Major Appliance
- Switching the Direction of a Ceiling Fan

Understanding Switches and Receptacles

- Replacing a Switch
- Replacing a Receptacle (Outlet)

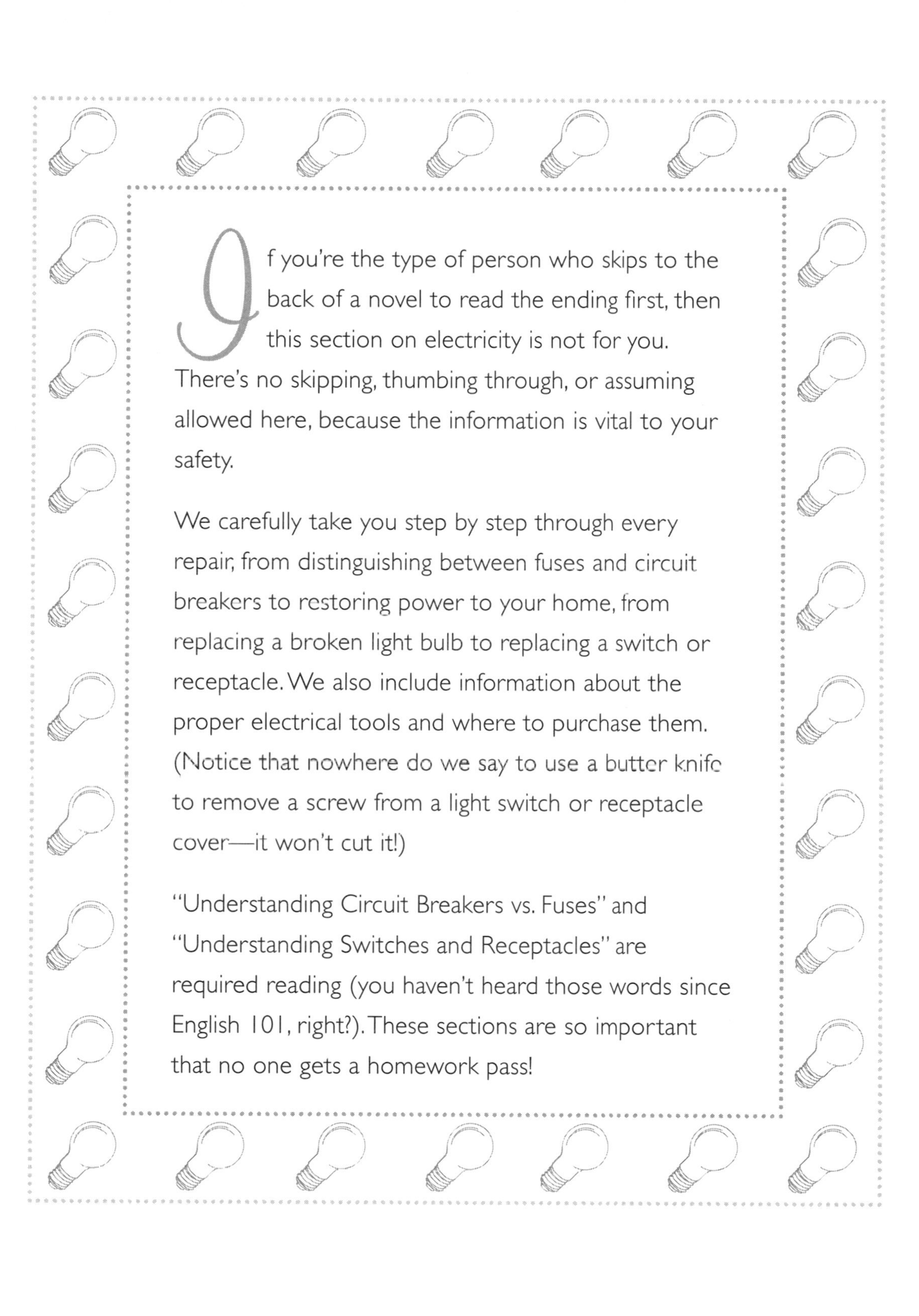

If you're the type of person who skips to the back of a novel to read the ending first, then this section on electricity is not for you. There's no skipping, thumbing through, or assuming allowed here, because the information is vital to your safety.

We carefully take you step by step through every repair, from distinguishing between fuses and circuit breakers to restoring power to your home, from replacing a broken light bulb to replacing a switch or receptacle. We also include information about the proper electrical tools and where to purchase them. (Notice that nowhere do we say to use a butter knife to remove a screw from a light switch or receptacle cover—it won't cut it!)

"Understanding Circuit Breakers vs. Fuses" and "Understanding Switches and Receptacles" are required reading (you haven't heard those words since English 101, right?). These sections are so important that no one gets a homework pass!

Circuit Breakers vs. Fuses

Understanding Circuit Breakers and Fuses

As a newly divorced woman who had just purchased a home, Joan had a big learning curve ahead of her. She just didn't know how big until the electricity went off in her family room and she realized that she had never opened a main service panel in her life. Joan called an electrician, who took pity on her and gave her step-by-step instructions over the phone. After successfully flipping the circuit breaker switch, she was off and running on her own two feet, even if they were baby steps.

We know that a lot of you are new to learning about electricity, so we're going to take this slow and easy so you won't get *burnt out.*

Electricity doesn't magically appear in your home—it travels from the outside power line to the interior main service panel, from which it travels through branch circuits to receptacles, switches, and light sockets.

Every home, whether it's an apartment or a mansion, has either circuit breakers or fuses located in the main service panel (typically a metal box).

The purpose of circuit breakers and fuses is not to provide electricity to your home but rather to protect the wires they're connected to from overheating.

If you're not sure whether you have circuit breakers or fuses, take a peek inside the main service panel. If you see lots of switches (toggles), these are circuit breakers. If you see round glass-topped plugs or tubes, these are fuses. Homes that are 40 years old or more *and* have never had their electrical system updated will have fuses.

It's important to know that when working with a fuse or circuit breaker, you should use only one hand, while keeping the other at your side. The reason is that electricity will always find the quickest path to the ground. Therefore, if you have one hand on the main service panel and one on a fuse or circuit breaker, you could receive an electrical shock.

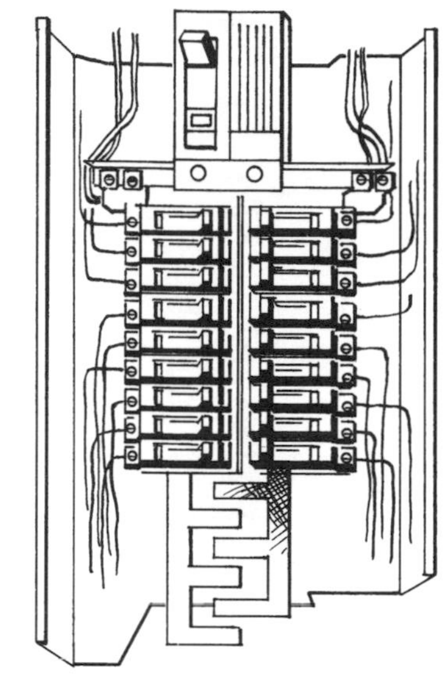

Main service panel (with circuit breakers)

Main service panel (with fuses)

Overloaded vs. Shorted Circuit

A circuit overloads when too many appliances or lights are using the same branch circuit. This is the most common reason for a fuse to blow or a circuit breaker to trip.

A short means that a bare wire has come in contact with another bare wire or metal object. This can happen in wall switches, outlets, and electrical cords or plugs and can result in a blown fuse or tripped circuit breaker.

Blown Fuse vs. Tripped Circuit Breaker

A blown fuse will appear discolored, or you will see that the lead strip inside has melted. In either case, the fuse will have to be replaced.

A circuit breaker that has tripped is easy to locate, because the toggle is in a position different from the others. Tripped circuit breakers don't need to be replaced, just reset. Call a certified electrician if a circuit breaker or fuse is hot to the touch, emits a buzzing sound, or is constantly blowing or tripping.

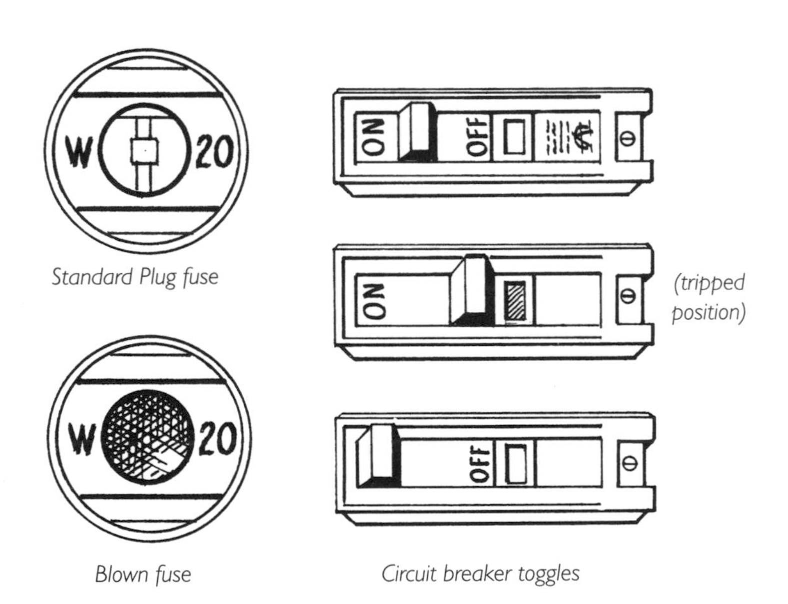

Standard Plug fuse

Blown fuse

Circuit breaker toggles

Creating a Circuit Map

When Diane received public housing after being on a waiting list for three years, she was grateful for a house her family could finally call *home*. After moving in, the kids turned on every light and appliance, causing an overload to the old circuitry. When Diane opened the main service panel and found that the circuits weren't identified, she realized the need to create a circuit map and enlisted the help of her children. The kids had so much fun they wanted to do it a couple of more times, which would have surely overloaded *Diane's* circuitry!

A circuit map, found on the inside door of the main service panel, provides a quick reference for locating specific circuit breakers or fuses. The map should list the circuit breakers or fuses for all the major electrical appliances and rooms served by each branch circuit.

If you think your electrician is the only one who would use the information on the circuit map, you're wrong. You need to refer to the circuit map when you're working on a receptacle or switch, and also in case of an emergency, because you need to quickly switch off the correct circuit breaker.

You'll only have to do this job once if your main service panel doesn't already have a circuit map, or if you're not the original homeowner. Depending on the size of your home, the job can be tedious, but if you have a friend willing to help you, it can be done much faster.

If you're using a radio, you'll need to turn it up loud enough so that you can hear it from the main service panel. Also, you'll have to unplug it and move it to a different location each time you test a circuit. Now you know why we suggested enlisting a helpful friend!

Turn *off* computers, televisions, and any other electrical equipment that's sensitive to power surges. Turn *on* the lights and ceiling fans in every room.

Before you open the door of the main service panel, make certain that the floor is completely dry to prevent electrical shock. Also, remember to always keep one hand at your side while using the other hand to remove a fuse or flip a circuit breaker (see "Understanding Circuit Breakers and Fuses," page 80).

Helpful friend relaying information

Look inside for a circuit map, and if there isn't one, make a list of all circuit breakers or fuses by giving each one a number with a blank space next to it. In this space, write the fixtures, appliances, and rooms serviced by that fuse or circuit breaker. Double toggles are designated for heat-producing appliances such as ovens, dryers, and heaters.

Because more than one circuit breaker can be assigned to a room, it's best to check each receptacle. (Aren't you glad you only have to do this once?)

No matter which fuse or circuit breaker you start with, it may take a few flips of toggles, or removal of fuses, before you hear the

Tools Needed

Helpful friend or plug-in radio

Pencil and paper

Walkie-talkies (helpful, but not necessary)

radio go off, or hear your friend yelling to you (this is where it's nice to have walkie-talkies). Write the name of the room and which appliances and fixtures went *off*.

Continue the process until every room and its main appliances and fixtures are coordinated with a circuit breaker or fuse. Place the new information inside the main service panel.

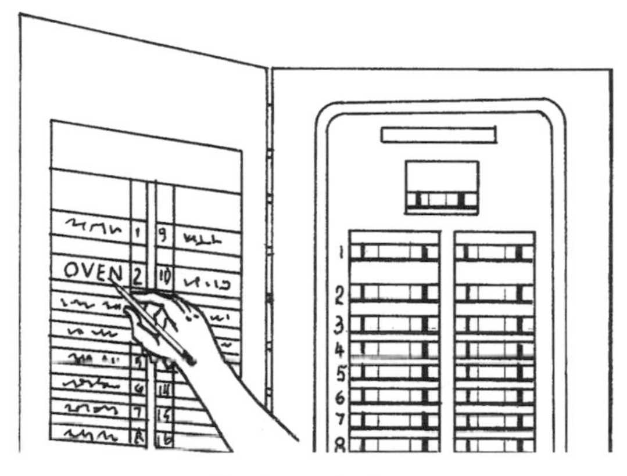

Creating a circuit map

Restoring Electrical Power

Did you know that hardwired smoke detectors won't work if your home's power is out? Only smoke detectors that are battery-operated or hardwired with a battery backup will work during a power outage, because they receive their power from the battery.

Raise your hand if you've ever had a major power outage and you didn't use the phone because you thought it wouldn't work? Don't worry . . . you're not the only one in the dark about electricity!

If you're completely without power, first check for a neighborhood outage. If everyone else has power but you don't, then your main breaker tripped, a fuse blew, or you just didn't pay your electric bill!

Those of you who live in an area where storms frequently down power lines could probably teach this section. You already know to call the electric company (never assume that someone else has), unplug all televisions and computers to protect them from a sudden electrical surge, test your home's Ground Fault Circuit Interrupters (see "Testing a Receptacle Ground Fault Circuit Interrupter (GFCI)," page 94), set the temperature in the refrigerator to its highest setting and keep the door(s) closed (it will cool off food faster when power is restored), and have candles and working flashlights with extra batteries on hand.

If your lights are out because you couldn't afford to pay your bill, there are federal, state, and local financial assistance programs specifically designed to help indigent people pay their energy bills. Contact your electric company, place of worship, or congressional representative or senator's office for more information. If you have the money to pay the electric bill but are ill or continually out of town, ask your electric company to set up a Third Party Notice program for you. This system allows the electric company to notify a person of your choice when your bill has not been paid.

If you have a vision problem, some electric companies will supply the bill in large print or Braille. Also, be sure to notify the electric company if you have electrically powered life-support equipment in your home so you can receive priority when power is restored.

Before beginning, make certain that the floor in front of the main service panel is completely dry to avoid electrical shock.

Remember that when working with a fuse or circuit breaker you should use only one hand, while keeping the other at your side. The reason for this is that electricity will always find the quickest path to the ground. Therefore, if you have one hand on the main service panel and one on a fuse or circuit breaker, you could receive an electrical shock.

If you have fuses, open the main service panel and use the flashlight to check for any that have blown. To find out if a fuse blew, look for one that is discolored or has a melted lead strip. Be sure to replace a blown fuse with one of the same amperage. If all the fuses are okay, then call your electric company.

If you have circuit breakers, open the main service panel and use the flashlight to check if the main circuit breaker, located at the top, has been tripped. If so, then reset it. A tripped circuit breaker will have its toggle in a position different from the others. If none of the breakers has tripped, call your electric company.

The telephone works in a power outage because the phone line is not connected to the power line that services your home's electrical system. However, it's best not to use the phone during an electrical storm because of the possibility of getting shocked.

Resetting a Circuit Breaker Switch (Toggle)

Does the thought of touching a circuit breaker switch make your hair stand on end? If learning about electricity has this effect on you, that's good, because you never want to take electricity too lightly (*hee hee*).

Tool Needed

Flashlight

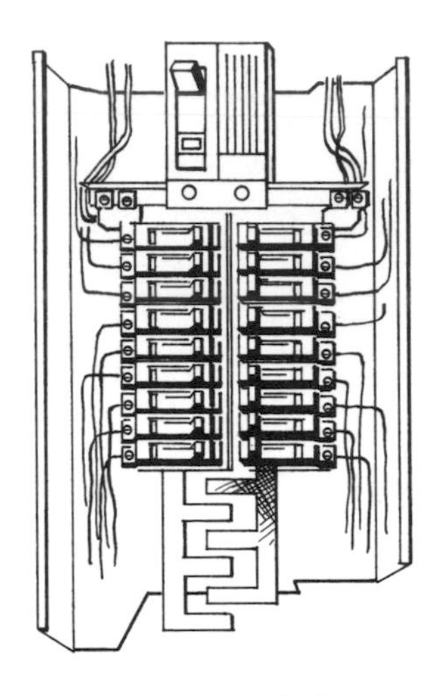

Main service panel (with circuit breakers)

First, unplug the electrical items you were using when the power went off. Before you open the main service panel, make certain that the floor is completely dry to prevent electrical shock. Use the flashlight to look inside the main service panel for the circuit breaker (a.k.a. toggle) that was tripped. Remember, a circuit breaker will trip because of a short in the circuit or an overload to the circuitry.

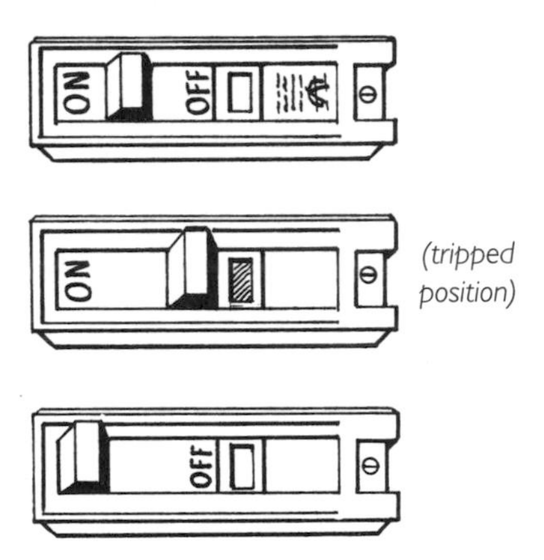

Circuit breaker toggles

Depending on the type of circuit breakers you have, the toggle may trip to the center or to the full *off* position. Before you begin resetting the circuit breaker, remember to keep one hand at your side, while using the other hand to touch the toggle (see "Understanding Circuit Breakers and Fuses," page 80). If the toggle trips to the full *off* position, reset it by flipping it to the *on* position. If it has tripped to the center position (some will show a small patch of red), reset by flipping the toggle to the *off* and then to the *on* position.

Plug in the appliances one at a time. If the circuit breaker trips again, then the problem could be that too many electrical items are using the same branch circuit. This means that you'll need to move some items to a different branch circuit. To do this, refer to the circuit breaker map located on the inside of the main service panel and note the electrical items delegated to the breaker that keeps tripping.

Another cause could be a short in the appliances. Check each electrical item and its cord and plug for damage. If any parts are defective, don't use the appliance again. Have it fixed or replace it.

If the circuit breaker keeps tripping after you've gone through the steps above, the problem lies within your home's electrical system, and you'll need to contact a certified electrician.

Replacing a Fuse

If you live in an older apartment or house, you know all too well the frustration of not being able to dry your hair while watching TV and toasting your bagel—all at the same time. Fuses can literally blow your morning routine!

Before you replace a fuse you need to know what kind of fuses you have and their amperage. Amperage (*amps* for short) is the unit of measurement for fuses. Fuses are color-coded and numbered according to their amps. For example, blue fuses are 15 amps, red/orange are 20 amps, and green fuses are 30 amps.

There are two types of household fuses plug and cartridge. A plug is a round glass-topped fuse that comes in three common styles: *standard, S-type,* and *time delay*. A *standard* plug screws into a socket in the fuse box like a light bulb. An *S-type* plug has two parts—the fuse and the adapter. The adapter goes into the socket first and then the fuse gets screwed into the adapter. This system is designed to keep

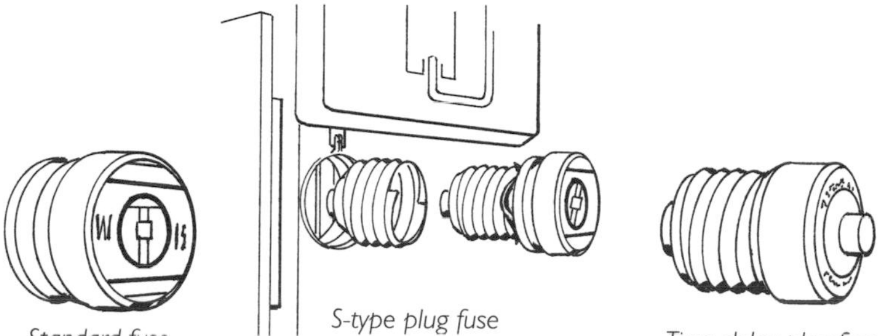

Standard fuse

S-type plug fuse

Time-delay plug fuse

Never replace a fuse with one of a different amperage. The larger amp will not provide you with more power, instead, it may cause an electrical fire.

Fuses, multimeters, and cartridge pullers can all be purchased at a hardware or electrical supply store.

someone from replacing a fuse with one of a different amperage. Once the adapter is in, it stays, but the fuse can be replaced. A *time-delay* plug prevents a sudden surge of electricity (e.g., an air conditioner turning on) from blowing the fuse.

Cartridge fuses are cylindrical and come in two different styles—*ferrule-contact* and *knife-blade-contact* (attractive names, huh?). The *ferrule-contact* fuse has rounded ends and is used for large appliances, such as ovens and clothes dryers. The *knife-blade-contact* fuse is used to protect the main electrical system. Never remove either of the cartridge fuses with your bare hands, because they could shock you. Instead, always use a plastic cartridge puller.

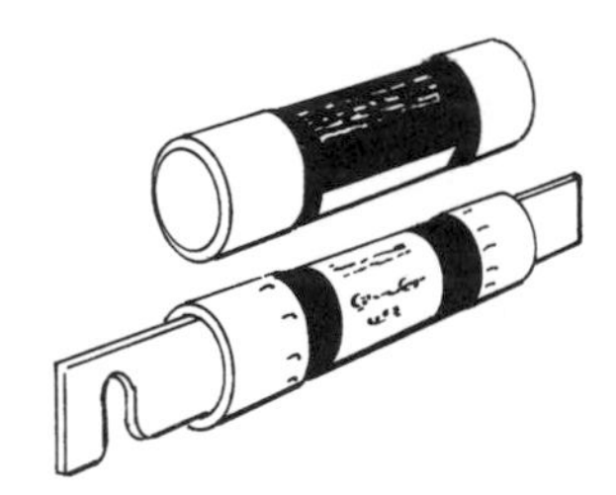

Cartridge fuses: Ferrule-contact and knife-blade-contact plug fuse

Replacing a Plug Fuse

First, unplug the appliance(s) you were using when the power went off. Before you open the main service panel, make certain that the floor is completely dry to prevent electrical shock. Use the flashlight to look inside the main service panel for the fuse that is discolored or has a

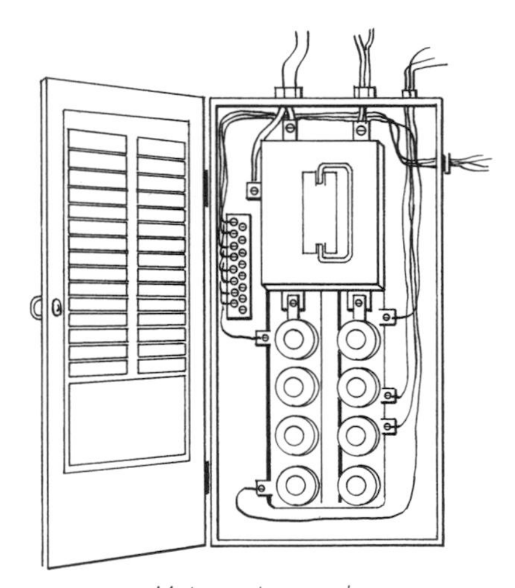

Main service panel

Tools Needed

Flashlight

New fuse (if applicable) 

melted lead strip. If a fuse is discolored, a short caused it; if the metal strip inside the fuse is melted but the glass is clear, there was an overload.

Standard plug fuse

Blown fuse

If there was an overload, it was caused by too many electrical items using the same branch circuit. Therefore, move some items to another, less burdened branch circuit before replacing the blown fuse. To do this, refer to the circuit map found on the inside of the main service panel. Note the fuse that blew and which electrical items are delegated to that fuse.

An electrical appliance can short when its plug or cord is damaged, so check for any defective parts and don't use it again. Have it fixed or replace it.

Before removing the fuse, remember to always keep one hand at your side, while using the other hand to touch the fuse (see "Understanding Circuit Breakers and Fuses," page 80). Turn the fuse counterclockwise to remove it. Place the new fuse (of the same amperage) into the socket and turn it clockwise.

If the fuse keeps blowing after you've gone through the steps above, then the problem lies within your home's electrical system, and you'll need to contact a certified electrician. A reputable electrician is likely to tell you your home's electrical system needs to be updated to circuit breakers for safety reasons, and it's probably true.

Replacing a Cartridge Fuse

First, unplug the electrical items you were using when the power went off. Before you open the main service panel, make certain that the floor is completely dry to prevent electrical shock. Unlike with plug fuses, when you open up the main service panel you won't be able to see which cartridge fuse has failed. The only way to check for a faulty cartridge fuse is by removing it with a cartridge fuse puller and testing with a multimeter.

Use the flashlight to look inside the fuse box. Cartridges are housed either individually in the service panel or in a fuse box that can be removed from the service panel. Before you take out a cartridge fuse, remember to keep one hand at your side, while using the other hand to hold the cartridge fuse puller (see "Understanding Circuit Breakers and Fuses," page 80).

To remove a cartridge fuse, place the head of the fuse puller around the middle of the fuse and pull it out. If the cartridge fuses are housed in a fuse box, use the lever to pull the box out first. *Never* touch the metal ends of the fuse, because they are very hot.

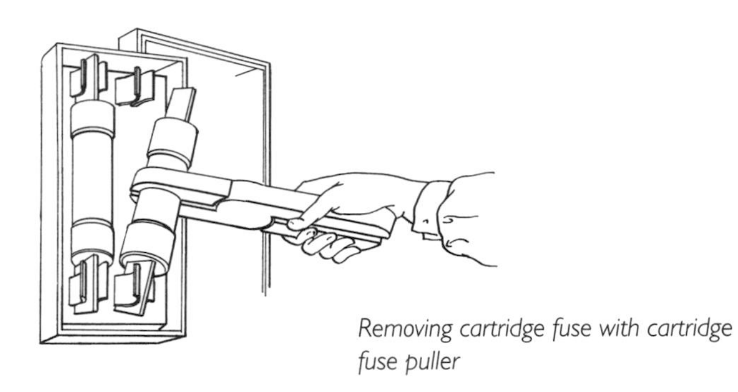

Removing cartridge fuse with cartridge fuse puller

Tools Needed

Flashlight 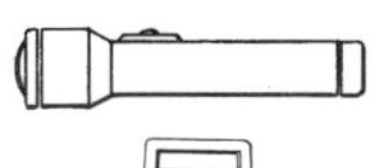

Cartridge fuse puller 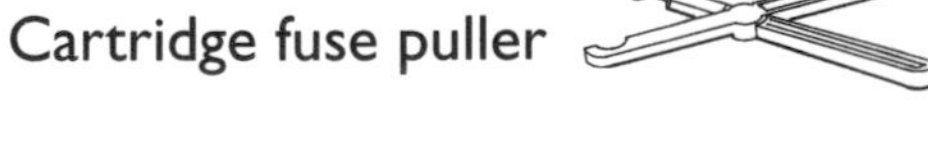

Multimeter 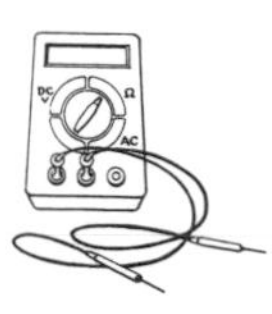

New cartridge fuse 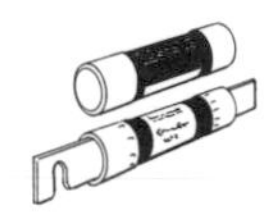

You'll need both hands to use the multimeter. This device allows you to see if the fuse can still receive electricity (i.e., if it's working properly). Set the multimeter to the RXI in the ohms range. Touch each end of the fuse with the probes. If the multimeter reads "0" ohms, the fuse works; if it doesn't, you need to replace the fuse. You don't need to use the plastic fuse puller when you replace the fuse.

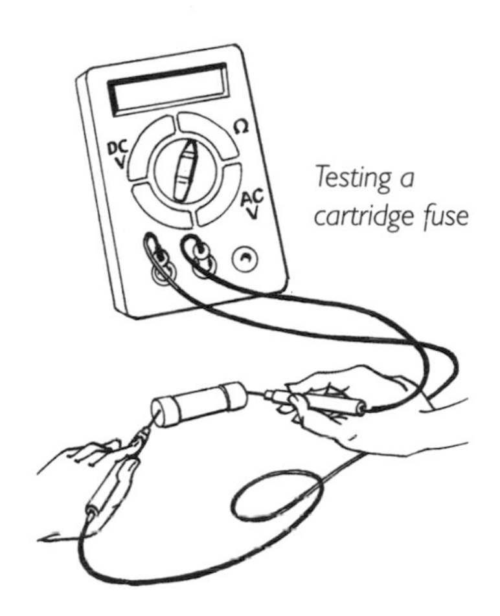

Testing a cartridge fuse

If all of the cartridge fuses tested positive, the problem lies within your home's electrical system, and you'll need to contact a certified electrician.

Testing a Receptacle Ground Fault Circuit Interrupter (GFCI)

FILL IN THE BLANK: You're in the kitchen washing a sink full of dirty dishes next to your daughter who's whipping up a batch of brownies, when all of a sudden she gets butter fingers and knocks the mixer into the sink. You're alive and well because of a properly working ________________.

(ANSWER: GROUND FAULT CIRCUIT INTERRUPTER.)

Never use electrical appliances near a sink or bathtub.

Some new hairdryers come equipped with a GFCI installed in the plug to protect you from electrical shock.

Don't be embarrassed if you don't know what Ground Fault Circuit Interrupters (GFCI) are—most people have never heard of them. GFCIs are designed to shut off the electricity emitting from a receptacle (outlet) before you get shocked or electrocuted. In other words, they help protect us from ourselves! They work by detecting "leaks" in the electrical current in a circuit branch. If a GFCI detects even minute differences between the amount of current flowing into and out of the receptacle, it immediately cuts power to the receptacle.

How can you tell if your home has GFCIs? Homes built since the 1970s typically have at least one GFCI but, to make sure, look for test and reset buttons on the receptacles in the bathrooms, laundry room, and kitchen. Also, look inside the main service panel for a GFCI circuit breaker. If you don't see any GFCIs in your home, contact a certified electrician to have them installed.

There are three types of GFCIs: 1) *receptacle*; 2) *plug-in*; and 3) *circuit breaker*. The receptacle GFCI can be found in areas where electricity and water are used in close proximity, such as kitchens, bathrooms, and laundry rooms. Plug-in GFCIs, which fit into a common receptacle, are effective as a temporary solution. These are commonly used for outdoor electrical devices such as garden trimmers and lawn mowers. Circuit breaker GFCIs are found in the main electrical service panel and replace the standard circuit breaker. This type of GFCI protects all receptacles on the circuit branch.

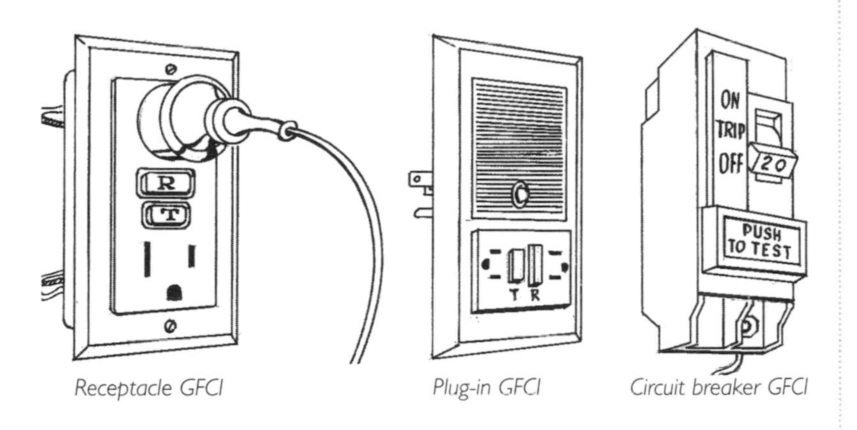

Receptacle GFCI Plug-in GFCI Circuit breaker GFCI

It's critical to your safety that you test your GFCIs monthly because they can become defective over time, and after every electrical storm, especially if you live in an area where there are frequent thunderstorms.

Push the reset button ("R") on the receptacle GFCI. Plug in the lamp and turn it *on*. Now push the test button ("T") on the receptacle GFCI. The light should turn *off*. Once again, push the reset button. The light should turn *on*.

If the GFCI failed this test, it may need rewiring, or it may have been improperly installed. In either case, don't rely on the GFCI to protect you from shock. Call a certified electrician to inspect the receptacle.

If an appliance such as a hairdryer or mixer falls into water, DO NOT TOUCH it. Immediately go to the main service panel and shut *off* the power to that circuit branch. Then use a voltage tester, which detects electricity, to verify there is no electrical power emitting from the receptacle the appliance is plugged in to (see "Understanding Switches and Receptacles," page 102). Once you've determined the power has been completely shut off to the receptacle, unplug the appliance and remove it from the water.

Tool Needed

Small lamp or nightlight

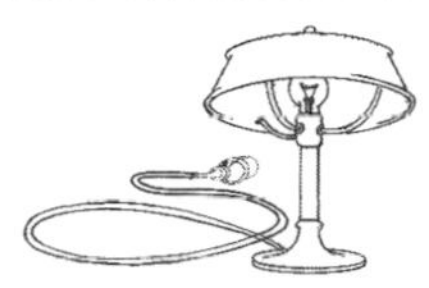

Light Bulbs

Removing a Broken Bulb from a Light Fixture

The correct wattage for the light fixture is typically imprinted either inside or on the exterior of the socket. If you don't see a number, never use a light bulb in that fixture that exceeds 60 watts, because it may cause an electrical fire.

It's important to use safety goggles when doing this repair because the glass from the light bulb can shatter.

Whether you say *po-ta-to* or *po-tah-to* or spell it with an "e," the spud is the best tool for removing a broken light bulb.

If the broken light bulb is in a ceiling fixture, turn *off* the electrical power by removing the fuse or flipping the circuit breaker for that light switch. If the broken light bulb is in a lamp, turn *off* the switch and unplug it from the receptacle.

Cut the potato in half. Wearing goggles, carefully push the white part of the potato into the broken bulb. Turn the potato counterclockwise until the bulb is completely unscrewed from the socket. If the bulb won't budge, use the needle-nose pliers.

Inserting potato into broken light bulb

Replace it with a new light bulb of the correct wattage. Restore the power at the main service panel, or plug the light fixture into the receptacle.

Tools Needed

Raw potato 

Needlenose pliers 

New light bulb

Safety goggles 

Replacing a Light Bulb in a Major Appliance

Are you wearing night-vision goggles for those midnight food raids? At ease, *woman*. Replace the appliance light bulb right away. And that's an order!

Light bulbs in refrigerators, ovens, and dryers, called "appliance bulbs," are typically 40 watts. In a refrigerator, the bulb is either in plain view or behind a cover. An oven light is always behind a cover or a wire protector. Not all clothes dryers have light bulbs, but if yours does, the bulb is probably 10 watts and can be found behind a protective case. All of these bulbs can be purchased in hardware and electrical supply stores and should be replaced only with bulbs of the correct wattage.

Appliances come in a wide range (*hee hee*) of styles, so you may need a small adjustable wrench to remove the appliance bulb cover, while your neighbor's appliance may require a screwdriver. Therefore, it's important for you to consult your appliance owner's manual before beginning, and if you have any questions, to call the manufacturer.

Tools Needed

Owner's manual 

Appliance light bulb (10, 20, or 40 watts) 

Phillips screwdriver 

Adjustable wrench (if necessary) 

Rubber gloves 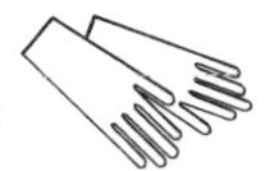

Replacing a Refrigerator Bulb

Before beginning, either unplug the refrigerator or turn *off* its power at the main service panel to avoid electrical shock. Refer to the owner's manual to correctly take *off* the cover. Remove the bulb and replace it with a new one. Return the cover and restore power to the refrigerator.

Replacing an Oven or Clothes Dryer Bulb

Before replacing a light bulb in an oven or dryer, turn *off* the electrical power at the main service panel to avoid electrical shock, and let the bulb cool before removing. If the light bulb is not behind a cover, unscrew the bulb and replace it with a new one of the same wattage.

Replacing light bulb in oven

If the cover for the light bulb is a wire protector, pull down on the wire to release it and remove the bulb. Replace the bulb and return the wire protector. If the light bulb is covered with a glass protector, use the Phillips screwdriver to remove the screws from the cover. Take the cover off, unscrew the bulb, and replace it with a new one. Return the cover and screws.

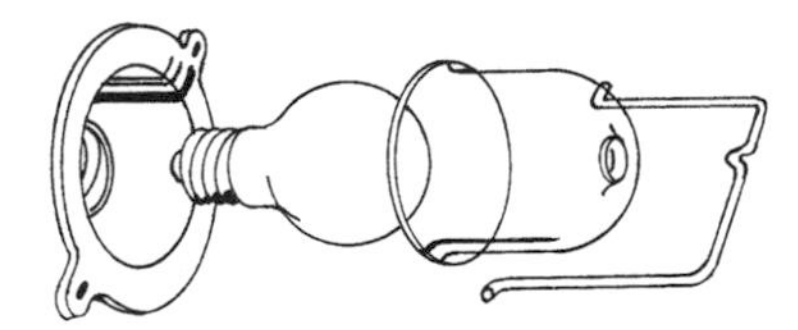

Disassembled appliance light bulb

If your oven is very dirty and the light bulb has never been replaced, the bulb may be greasy. If so, wear the rubber gloves to provide a better grip. If there is a lot of grease build up, you may have a difficult time removing the cover and/or wire protector. Be careful not to pull too hard on the wire protector, cover, or bulb because the entire encasement could come out, exposing the wires. (This is why we advised shutting *off* the power to the oven at the main service panel—for protection against shock or an electrical fire. Call the appliance manufacturer for a listing of service professionals in your area and then call out for pizza, because you won't be cooking until that problem gets resolved . . . that is, if you want it fixed!

Switching the Direction of a Ceiling Fan

If you can't stand the heat, don't get out of the kitchen. Maybe all you need to do is switch the direction of your ceiling fan blades!

Ceiling fans have stood the test of time for providing an efficient and effective way to circulate air in the home. In warm weather a ceiling fan's blades should rotate in a counterclockwise direction to draw the hot air upward. In cool weather, the fan blades should rotate clockwise to force the warm air downward. How can ceiling fans do both? It's as easy as flipping a switch!

Turn the ceiling fan *on* to watch which direction the blades rotate. Remember, clockwise for cool weather, counterclockwise for warm weather. Turn the ceiling fan *off*.

Use a stepladder or ladder to reach the still ceiling fan. Move the slide switch on the head of the fan to the opposite direction. Step away from the fan before turning it *on* (*duh!*).

Tool Needed

Stepladder or ladder

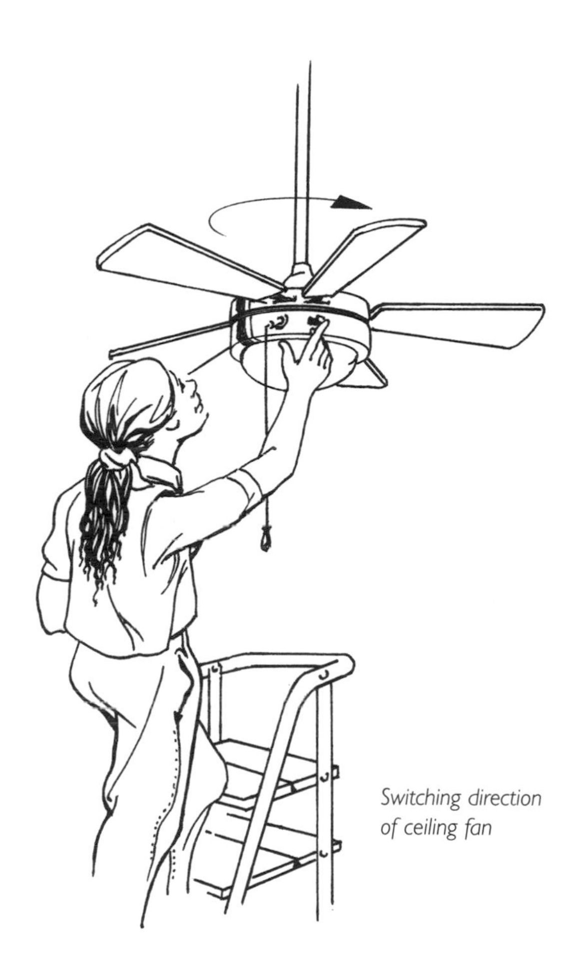
Switching direction of ceiling fan

Even if you have a remote control that operates your ceiling fan, it may not have a reverse feature. Therefore, the only way to change the direction of the blades is to do it manually.

Understanding Switches and Receptacles

We wish we could tell you that each of you will have the same type of wiring in your home, but we can't. Some of you live in old homes where the wiring has never been updated, while some live in homes where an electrician went haywire with the electrical system. Therefore, we're sticking to information about standard circuits, basic switches, and receptacles. If you find that the wires in your home differ from what is standard, stop and call a certified electrician.

All About Wires

Behind every switch and receptacle cover is an electrical box (plastic or metal) that houses the wires that attach to a receptacle or switch.

Wires are covered with insulation to prevent electrical shocks and are distinguished by a color code for standard circuits: black is always a hot wire, white is always a neutral wire, and green is always a grounding wire.

An older home whose electrical system has not been updated will *not* have a grounding wire (green or bare copper) and will have one black and one white wire. In this situation, the white wire may work as a hot wire. If done properly, an electrician will have wrapped a piece of black electrical tape to indicate that it's a hot wire.

A wire will never be completely covered with insulation because you need to be able to wrap the exposed wire around the terminals (screws), located on the sides of a switch or receptacle. An older home, whose electrical wiring has never been updated, may have aluminum instead of copper wiring. If you have aluminum wiring, have a certified electrician do any repairs, instead of doing them yourself.

Wires are either attached to the terminals on the side of the switch or receptacle or inserted into holes in the back called "backwire push-in connections." The Consumer Product Safety Commission recommends against back-wire push-in connections because the wires can become loose and overheat. The safest way (and the only way we're going to show you) is to reconnect the wires around the terminals (screws) of a switch or receptacle.

Testing for Power

When you test a switch or receptacle to see if you shut its power *off* at the main service panel, you have to check it more then once.

To test for power at the switch and receptacle we chose a voltage indicator because of its simple design, ease of use, and accuracy. It can be purchased in electrical supply or hardware stores.

Voltage indicator

Before using a voltage indicator, you need to verify that it's working properly. First, check the batteries by rubbing the tip of the voltage indicator vigorously against your shirtsleeve or pants to create static. It will emit a continuous beep and flashing light when it detects electricity.

Next, place the tip of the voltage indicator in the hot side (smaller or right-hand slot) of a receptacle that you know works. If it beeps and its light flashes, then it's functioning properly.

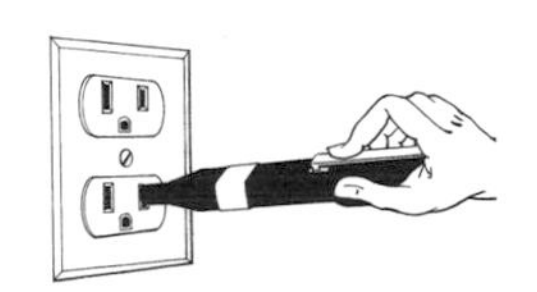

Inserting voltage indicator into hot slot

We need to warn you that an electrician can make mistakes, as can a previous homeowner who may have done some electrical wiring. Therefore, have a certified electrician check your home's wiring once every ten years, especially if you've added high-wattage appliances, and after you've purchased a home.

Replacing a Switch

No one actually says, "Don't forget to turn off the single-pole switch, honey," or, "Dear, did you leave the three-way switch on again?" A light switch is a light switch, right? Well, not really.

The difference between a three-way and single-pole switch (besides the confusing names) is that the single-pole controls a light from only one location, whereas the three-way allows you to turn a light *on* and *off* from two locations. For example: If a room has one entrance, it probably has a single-pole switch. If it has two entrances, it will likely have a three-way switch so that you can turn on the same light no matter which way you enter the room.

Replacing a Single-Pole Switch

There are generally three reasons for replacing a single-pole switch: 1) you want to change its color; 2) you want to install a dimmer switch; or 3) the switch is broken. If you suspect that the switch is faulty, first rule out the light bulb, lamp, or circuit breaker/fuse as the source of the problem.

A new single-pole switch comes either packaged or loose in a box full of switches. Both types are UL approved, but the difference is that the packaged switch is of higher quality and costs more. Read the label on the box to verify that you're getting a single-pole switch or look on the switch for the words "on" and "off" and check to see if there are three terminals (screws). If there are four terminals and/or the words "on" and "off" are missing, you have a three-way switch in your hand.

Turn *off* the power at the main service panel. Using the screwdriver, loosen the screws on the switch plate cover and remove it (tape the screws to the cover so you won't lose them).

Removing the Old Switch

Test to see if the electricity to the switch was completely shut *off* by touching the wires and terminals (screws located on the sides of the switch) with the voltage indicator (see "Understanding Switches and Receptacles," page 102). If it emits a continuous beep and flashing light, then the electricity was **not** shut *off*. Go back to the main service panel and do it right this time!

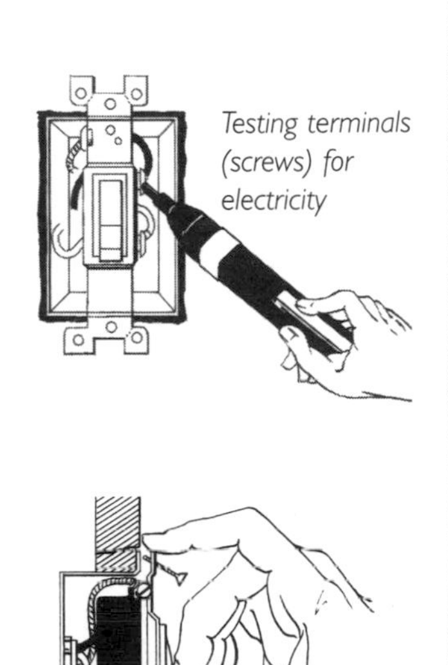

Testing terminals (screws) for electricity

Remove the screws that attach the switch to the box. The switch is not completely free yet because the wires are still attached. Hold the switch and gently pull it out of the box so you'll have more room to work. Identify the wires with masking tape to make it easier when reassembling.

Pulling out the old switch

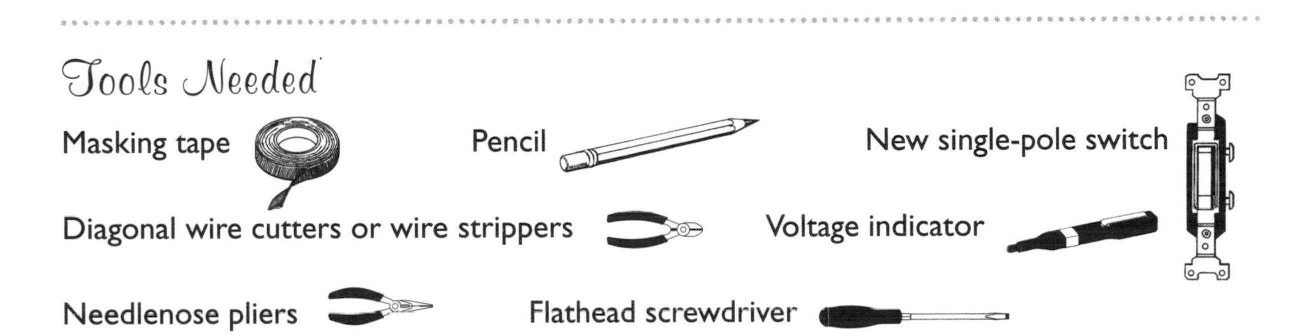

Tools Needed

Masking tape

Pencil

New single-pole switch

Diagonal wire cutters or wire strippers

Voltage indicator

Needlenose pliers

Flathead screwdriver

Use the screwdriver to loosen the terminals on the switch and unwind the wires from around them. Or, if the wires were inserted into the holes in the back of the switch, place a small screwdriver (a metal nail will work, too) in the slot located next to each hole to release the wires. If this doesn't work, use the wire cutters to snip off the wires as close to the switch as possible. Remove the old switch.

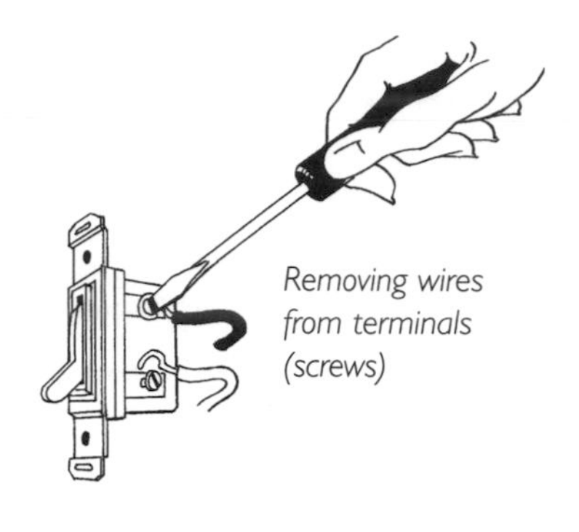

Removing wires from terminals (screws)

Installing a New Switch

You'll need about ½ inch of bare wire to wrap around a terminal, so you may have to cut off some of the wire's insulation with the wire cutter. Pull off the insulation with the wire cutter or needlenose pliers.

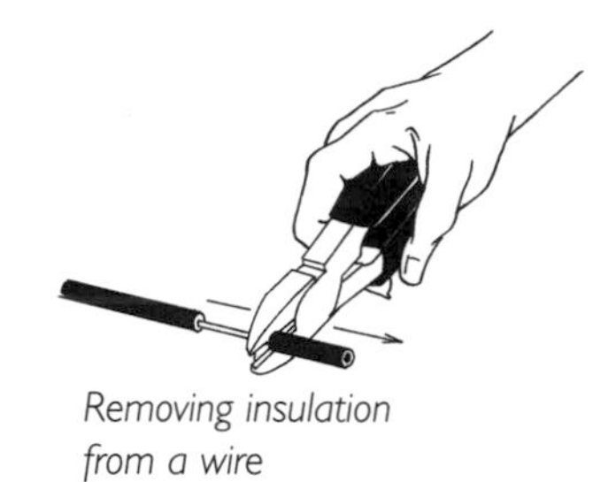

Removing insulation from a wire

The two insulated wires you removed from the single-pole switch will either both be black or one black and one white with a black marking (indicating that it's a hot wire). These wires should be fastened around the two metal terminals with no difference in the order they're installed because they're both hot wires. You may also have removed a green insulated wire or a bare copper wire. This is the grounding wire, which can be connected only around the green terminal on the switch.

Hold the new switch in one hand, making sure the word "off" is at the top. This ensures that you will install the switch in the correct position. If it reads "no" or "ffo" the switch is upside down.

Bend the tip of each wire into a loop with the needlenose pliers and curl it around its terminal (as marked by the masking tape) in a clockwise direction. Using the screwdriver, tighten the terminal into place. A good safety practice is always to connect the bare copper or green (grounding) wire first before you connect the other wires.

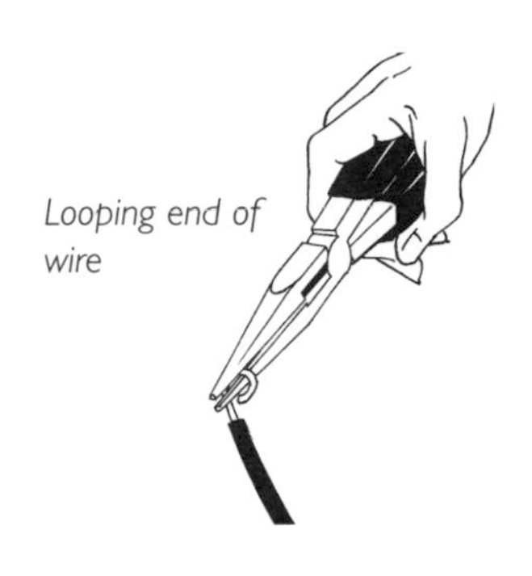
Looping end of wire

When all of the wires have been properly connected, carefully push the switch back into the box, making sure that the grounding wire is not touching the other wires (you may have to use the pliers to gently move it away). Turn *on* the power to the switch at the main service panel. Using the voltage indicator, check that the switch is receiving electricity. If it isn't, turn *off* the power at the main service panel and go through the steps again. If the switch still doesn't work, it's time to call a certified electrician.

Replacing a Single-Pole Switch with a Dimmer Switch

A dimmer switch allows you to adjust the degree of lighting in a room, as well as to turn the light *on* and *off*.

Dimmer switches come in many different colors and styles, but

Tools Needed

Flathead screwdriver 

Voltage indicator 

Diagonal wire cutters or wire strippers 

Dimmer switch (including two screws and three caps) 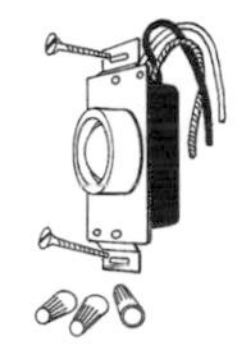

the most important thing to check when purchasing is what type of switch—single-pole, three way, etc.—it's replacing (remember, single-pole for this repair). This information can be found on the package. Also be aware that the cover to a dimmer switch is typically sold separately.

Turn *off* the power at the main service panel. Using the screwdriver, loosen the screws on the switch plate cover and remove it (tape the screws to the cover so you won't lose them).

Removing an Old Switch

Test to see if the electricity to the switch was completely shut *off* by touching the wires and terminals (screws located on the sides of the switch) with the voltage indicator (see "Understanding Switches and Receptacles," page 102). If it emits a continuous beep and flashing light, the electricity was **not** shut *off*. Go back to the main service panel and do it right this time!

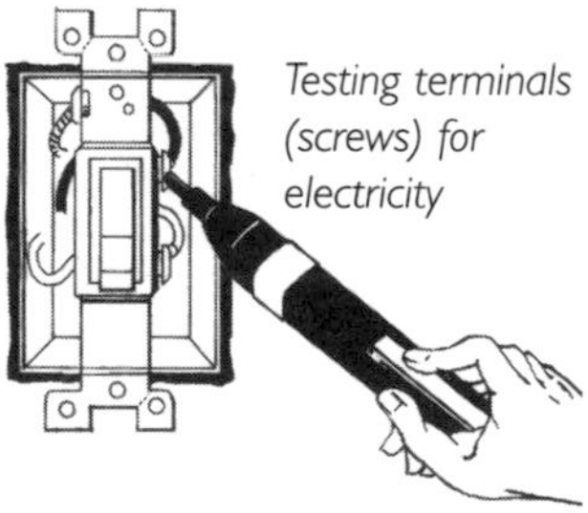

Testing terminals (screws) for electricity

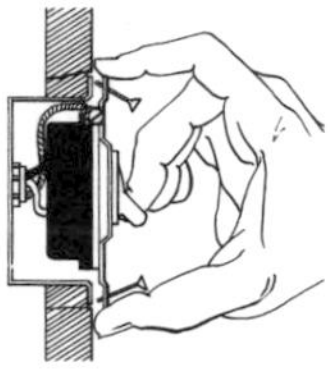

Remove the screws that attach the switch to the box. The switch is not completely free yet because the wires are still attached. Hold the switch and gently pull it out of the box so you'll have more room to work. Use the screwdriver to loosen the terminals (screws) on the switch and unwind the wires from around them. Remove the old switch.

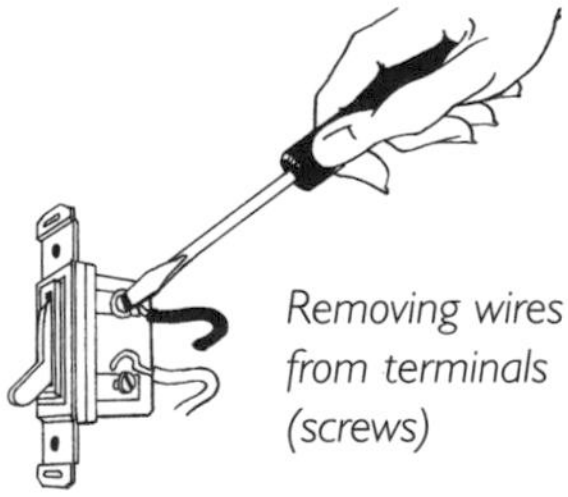

Pulling out the old switch

Removing wires from terminals (screws)

If the wires were inserted into the holes in the back of the switch, place a small screwdriver (a metal nail will work well, too) in the slot located next to each hole to release the wire. If this doesn't work, use the wire cutters to snip off the wire as close to the switch as possible. Remove the old switch.

Installing a New Dimmer Switch

Join the wires from the box that were originally attached to the single-pole switch with the wires from the dimmer switch. You'll need about ½ inch of bare wire to connect with another wire, so you may have to cut some of the wire's insulation with the wire cutter. Pull off the insulation with the wire cutter or needlenose pliers.

First connect the green or bare copper (grounding) wires by placing them side by side in your hand and slightly twisting the tops of the wires together. Place a cap (a.k.a. nut) on top, covering both wires, and screw the cap clockwise until it's tight. Next, pick one of the black (hot) wires from the dimmer switch and connect it with a black wire from the box, slightly twisting the tops of the wires together. Place a cap on the top, covering both wires, and screw the cap clockwise until it's tight. Repeat the same steps for the white (neutral) wires. Be careful to check that all of the caps are tightened and that no bare wires are exposed.

Gently push the wires into the box and insert the mounting screws. Turn *on* the power to the switch at the main service panel. Using the voltage indicator, check that the switch is receiving electricity. If it is, replace the cover and screws.

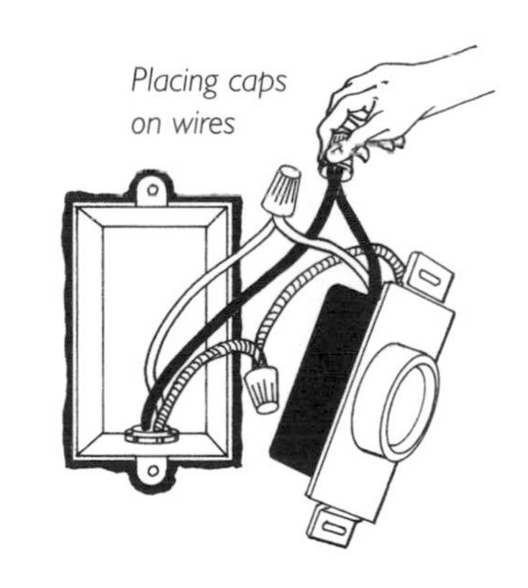
Placing caps on wires

If the new dimmer switch isn't working, turn *off* the power at the main service panel and go through the steps again. If the switch still doesn't work, it's time to call a certified electrician.

Replacing a Receptacle (Outlet)

A sign that a receptacle has developed problems, such as worn parts or loose wires, is that the cover is warm/hot to the touch. This signifies a potential electrical fire hazard, so contact a certified electrician immediately, and do not use that receptacle.

Don't let this electrical repair *jolt* you into a frenzy—instead, think of this project as earning you bragging rights at the office cooler or in the carpool line!

There are three types of common household receptacles: 1) two-pole polarized; 2) two-pole three-wire grounding; and 3) two-pole non-polarized. You're probably thinking, *I don't have those in my home!* but you do—you just didn't know they had names other than "outlet."

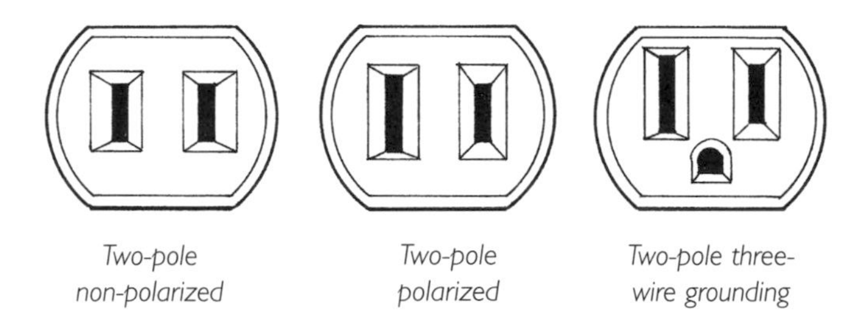

Two-pole non-polarized — *Two-pole polarized* — *Two-pole three-wire grounding*

Now it's time to learn about the innards of the receptacle. Read this section at least once while holding a new receptacle before you begin.

Identifying Receptacles

A new receptacle comes either packaged or loose in a box full of receptacles. Both are UL approved, but the difference is that the packaged receptacle is of higher quality and costs more. Make sure that the receptacle matches your old one in terms of it being polarized, non-polarized, and single or duplex.

Looking at the illustrations on page 110, you'll see that a two-pole non-polarized receptacle has two slots identical in size and shape. A two-pole polarized has one slot larger than the other, and a two-pole three-wire grounding has two slots (one larger than the other) and a half circle opening. Notice that receptacles can have one plug or two (one on top of the other), a "duplex."

One-plug receptacle

Duplex receptacle

When you're installing a two-pole polarized receptacle, the black (hot) wire should only be connected to the dark or brass terminal (screw) on the same side as the smaller slot, the white (neutral) wire should only be connected to the light or silver terminal on the same side as the larger slot, and the green or bare copper (grounding) wire should only be connected to the green terminal. For a non-polarized receptacle (both slots are the same size), always use the voltage indicator (see "Understanding Switches and Receptacles," page 102) to find the hot slot.

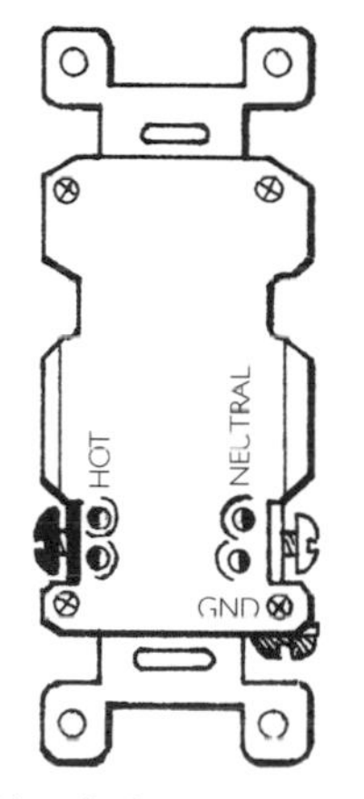

Identified terminals

Newer receptacles may carry letters or abbreviations on the side or back identifying hot, neutral, and grounding sites.

Testing a Receptacle

If you plug an appliance or fixture into a receptacle and it doesn't work, first check to see if the receptacle is controlled by a light switch

Tool Needed

Voltage indicator 

(you may need to flip the switch). Next, try plugging it into a different receptacle. If it works, then the trouble could be that a fuse *blew* or a GFCI or circuit breaker *tripped*. If neither of these things happened, the problem lies either with the receptacle or its circuit branch.

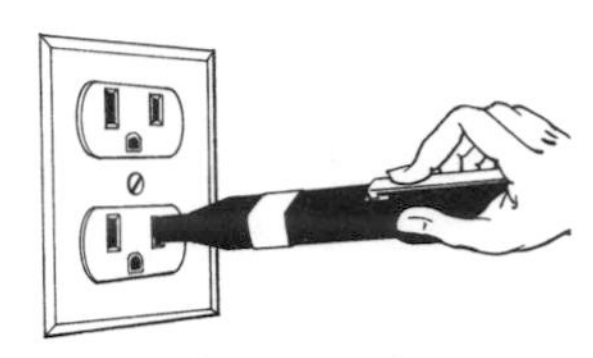

Inserting voltage indicator into hot slot

Put the tip of the voltage indicator into the small slot (hot) if it's a two-pole polarized or two-pole three-wire grounding receptacle; if it's a two-pole non-polarized receptacle, put the tip into the right-hand slot (see "Understanding Switches and Receptacles," page 102). If it emits a continuous beep and flashing light, then electricity is reaching the receptacle; therefore, the problem is with the fixture or appliance. If you have a duplex receptacle, test all four slots.

If there is no continuous flashing light and beep, a number of things could be wrong: 1) the receptacle is broken and needs to be replaced; 2) the receptacle is GFCI-protected and the reset button needs to be pressed; or 3) there's a problem with your home's electricity and you'll need to contact a certified electrician. If the problem doesn't lie with the GFCI, then you need to rule out a faulty receptacle.

Checking for a Faulty Receptacle

Turn *off* the power to the receptacle at the main service panel. Using the screwdriver, remove the screw and take off the cover (tape the screw to it so you won't lose it).

Tools Needed

Flathead screwdriver 

Masking tape 

Voltage indicator 

Use the voltage indicator to check all of the slots again for power. You need to test all of the slots, because the receptacle may not have been wired correctly (it's better to be cautious!). If it emits a continuous beeping sound and flashing light, the electricity was **not** turned *off*. Go back to the main service panel and do it right this time!

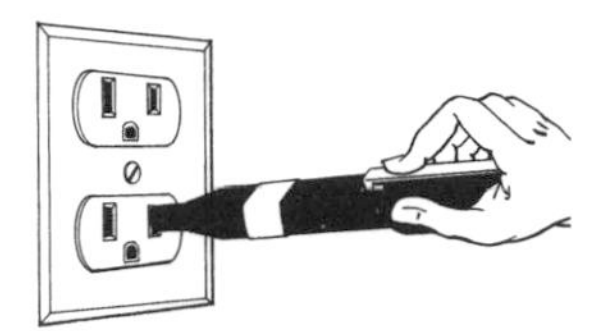

Inserting voltage indicator into hot slot

Remove the screws on the top and bottom that hold the receptacle to the box in the wall. Gently pull the receptacle out. Turn *on* the power to the receptacle at the main service panel.

Pulling out receptacle

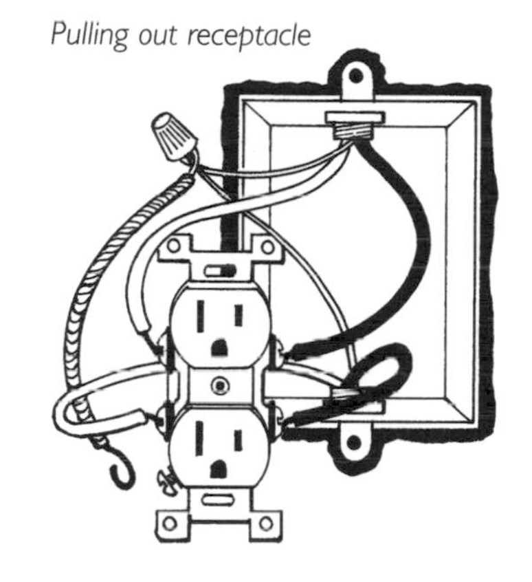

Locate the black wire (hot) and touch it with the tip of the voltage indicator (never touch a live wire with your hands). If the sensor beeps and the light flashes, the wire is receiving electricity, and the problem lies with the receptacle. You'll need to replace the receptacle, but don't worry, it's really not hard to do (see the following repair).

Testing the hot wire

If there are no signals, then electricity is *not* reaching the receptacle; therefore, the problem is with that circuit branch. Contact a certified electrician.

Turn *off* the power to the receptacle at the main service panel. Push the receptacle back into the box. Use the screwdriver to install the two screws in the receptacle box. Replace the cover and its screw. Turn *on* the power to the receptacle at the main service panel.

Replacing a Receptacle

Turn *off* the power to the receptacle at the main service panel. Using the flathead screwdriver, remove the screw and take off the receptacle cover (tape the screw to it so you won't lose it).

Using a voltage indicator, insert the tip into the small slot (hot) if it's a two-pole polarized or two-pole three-wire grounding receptacle. For a two-pole non-polarized receptacle, place the tip of the voltage indicator in the right-hand slot. You should test all of the slots, because the receptacle may not have been wired correctly (it's better to be cautious!). If the voltage indicator emits a continuous beeping sound and flashing light, then the electricity was **not** turned *off*. Go back to the main service panel and do it right this time!

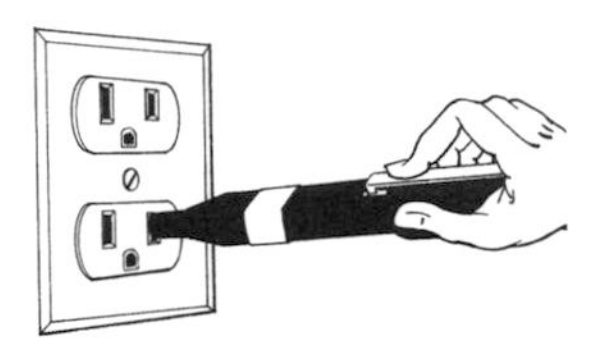

Inserting voltage indicator into hot slot

Removing an Old Receptacle

Using a voltage indicator, check one more time that the electricity is *off*. Remove the screws on the top and bottom that hold the receptacle to the electrical box in the wall. Gently pull the receptacle out.

Before removing any wires, mark each one with a piece of masking tape, noting the hot, neutral, and grounding wires.

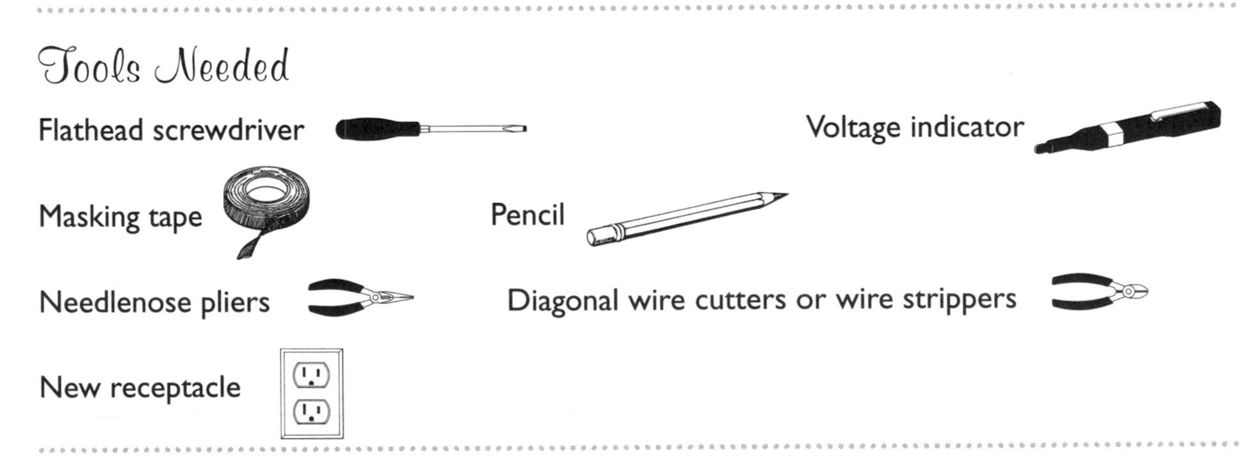

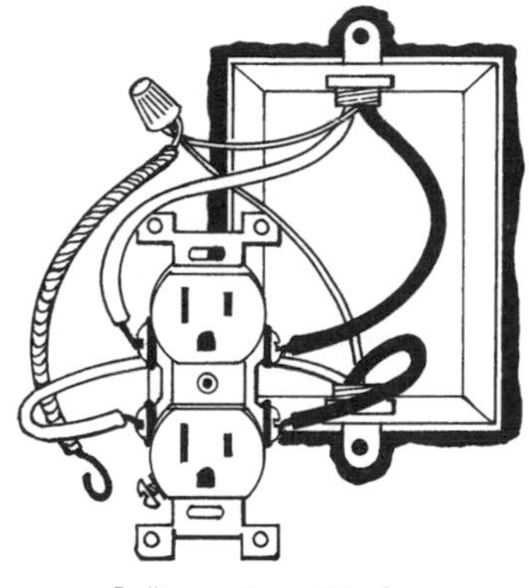
Pulling out receptacle

Identifying wires

Next, see if the wires are wrapped around terminals (screws) on the sides of the receptacle, or if they're inserted into holes in the back of it. If wires are attached around the side terminals, loosen the terminals with a screwdriver and detach the wires (if necessary with the help of needlenose pliers).

If the wires are inserted in holes in the back, place the small flathead screwdriver (a metal nail will work well, too) in the slot located next to each hole to release the wire. If this doesn't work, use the wire cutter to snip off the wire as close to the receptacle as possible.

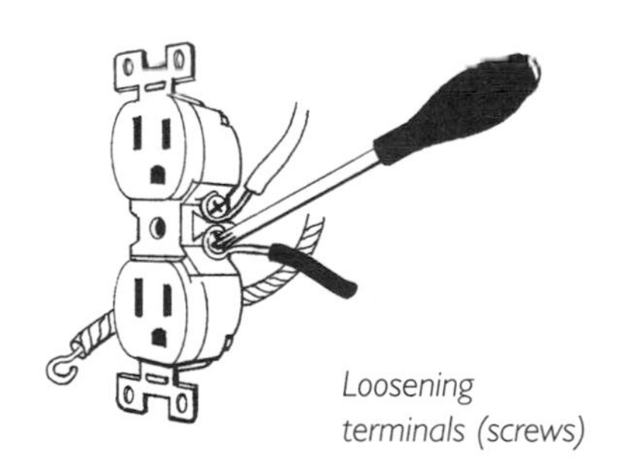
Loosening terminals (screws)

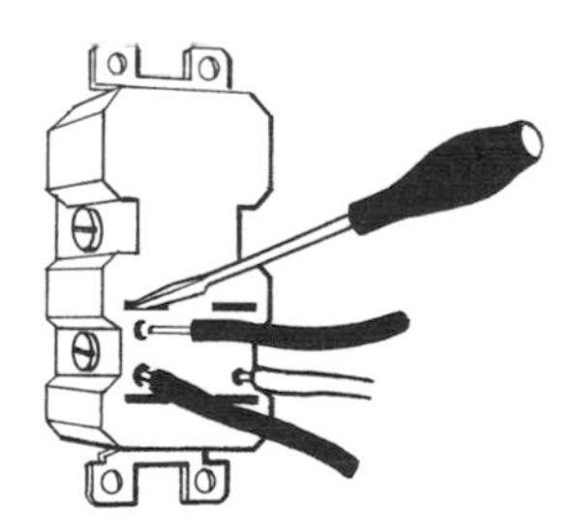
Releasing back wires

Installing a New Receptacle

Notice whether the old receptacle has a metal tab between the two terminals on each side. If a tab is missing on the hot side, it signifies that the receptacle is controlled by a light switch. Using wire cutters, snip the metal tab on the hot side of the *new* receptacle.

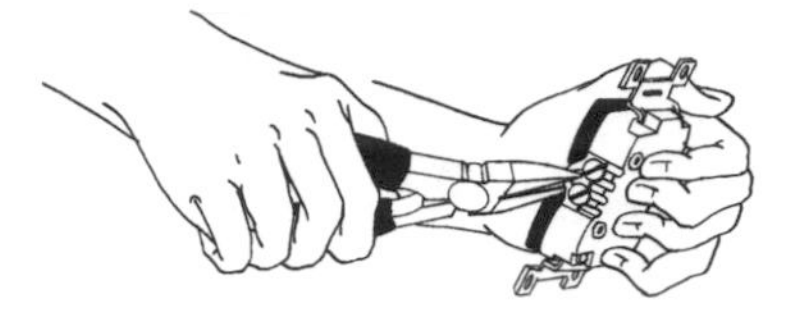

Snipping metal tab

You'll need about ½ inch of bare wire to wrap around a terminal, so you may need to cut off some of the insulation. Pull off the excess insulation with the wire cutters or needlenose pliers.

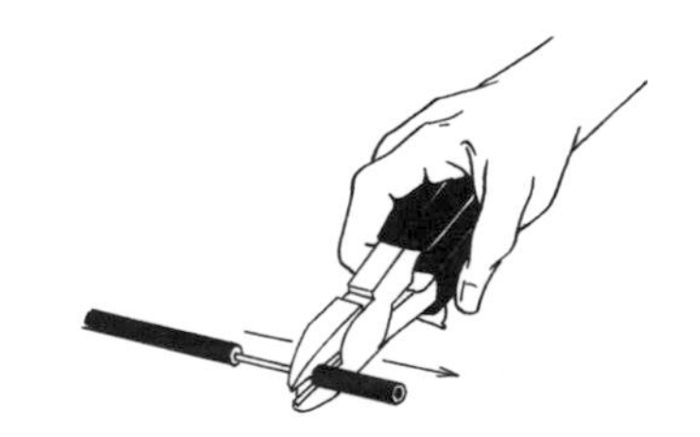

Removing insulation from a wire

It's always a good safety practice to connect the green or bare copper (grounding) wire first, then the white wire, and finally the black wire. Bend the tip of one wire into a loop with the needlenose pliers and curl it around the terminal in a clockwise direction. Using the screwdriver, tighten the terminal into place. Repeat the process with the other wires.

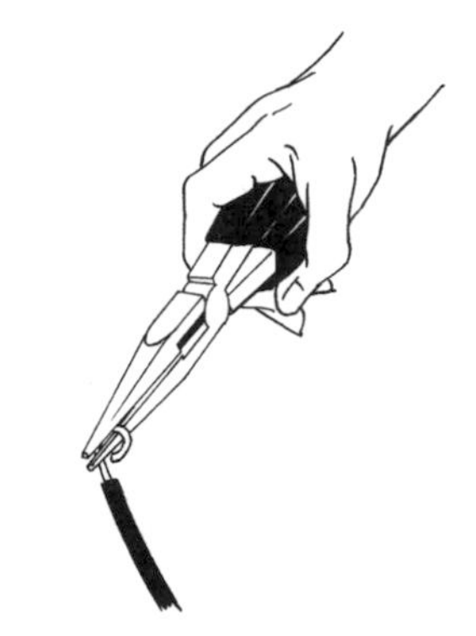

Looping end of wire

When all of the wires have been properly connected, carefully push the receptacle back into the electrical box, making sure that the grounding wire is not touching the other wires (you may have to use the needlenose pliers to gently move it away). Replace the screws connecting the receptacle to the electrical box.

Turn *on* the power to the receptacle at the main service panel. Use a voltage indicator to check the receptacle for power (see "Understanding Switches and Receptacles," page 102). If it works, *hooray!* Place the cover on it and tighten the screw. If it doesn't work, turn *off* the power at the main service panel and go through the steps again. Should the receptacle still not work, it's time to call a certified electrician.

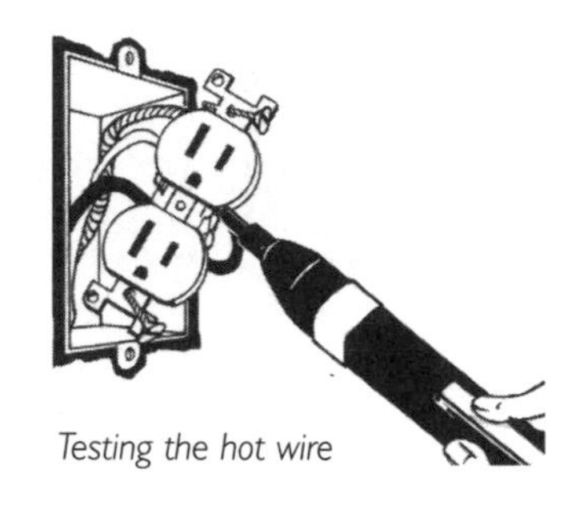
Testing the hot wire

Major Appliances

- Painting Touch-Ups for Appliances
- Leveling Appliances
- Freeing a Jammed Garbage Disposal
- Fixing a Dishwasher
- Maintaining a Refrigerator
- Cleaning a Stove's Grease Filter
- Repairing a Slow-Filling Washing Machine
- Cleaning a Clothes Dryer Hose and Vents
- Cleaning or Replacing a Furnace Filter
- Lighting a Pilot in a Gas Furnace

Our definition of a major appliance is something designed for household use that's cheaper to fix than replace, such as a washing machine, clothes dryer, refrigerator, oven, dishwasher, and furnace. There are a lot of things that can go wrong with an appliance, but the repairs included in this section are ones you can easily do yourself.

When in doubt, the best tools for fixing a major appliance are your phone and owner's manual. If you're missing the owner's manual, request a copy from the manufacturer. The leading appliance manufacturers have customer hotlines staffed with helpful assistants who guide you in assessing the problem, explain how to fix it, supply you with certified contractors in your area, or order parts. They also have websites that come in quite handy, too.

Please note that electrical safety rules apply to this section as well.

Painting Touch-Ups for Appliances

If you can paint your nails, you can do this repair. Just don't get the enamels mixed up!

If an appliance has chipped, call the manufacturer for the correct color. You'll need to provide the appliance's serial and model numbers because some manufacturers can have seven different shades of "white" for their appliance line. Once you have the exact color code, you have the option of purchasing the paint directly from the manufacturer or from a home appliance store. Before starting this project, be sure to read the manufacturer's label, and be aware that the little bottle of paint packs a powerful fume.

Applying the Paint

Before you begin, make sure that the area is free of dirt. Paint the chipped spot as you would paint your nails—apply a thin layer. Wait one hour before applying another coat, if necessary.

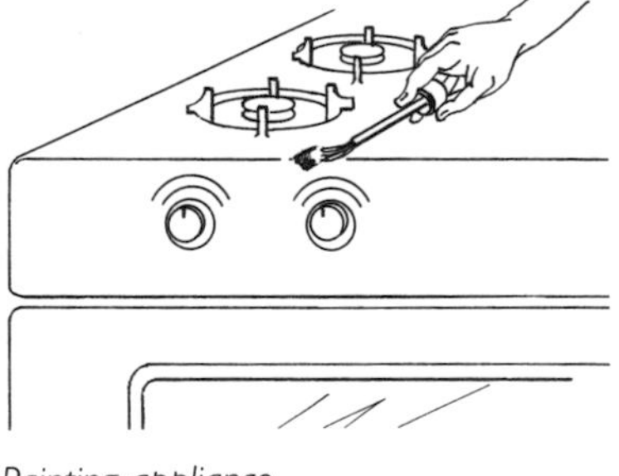

Painting appliance

Tool Needed

Appliance touch-up paint (includes brush)

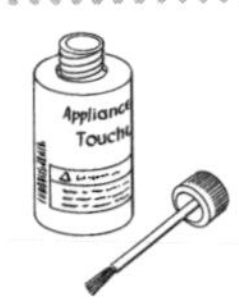

Leveling Appliances

Marianne's washing machine would do the two-step when it was on, but she put off fixing it because when she placed her colicky baby in a carrier on top, the vibrations rocked her to sleep. She was so desperate for her baby not to cry that she ended up begging to do all her neighbors' laundry!

In Marianne's case, we say: whatever works! But to keep your own appliances running smoothly, you'll need to remedy the problem as soon as possible.

All appliances are leveled when they are professionally installed; however, over time they can become unbalanced because of vibration. The two appliances most commonly associated with balance problems are washing machines and refrigerators.

An appliance will vibrate more than normal when the feet of the machine are not level. Because appliances differ, read and follow the manufacturer's instructions before beginning this project.

Refrigerator

Refrigerators are designed to tilt backward so the doors will close easily. If the appliance is unbalanced, the door won't close properly,

Carpenter's level 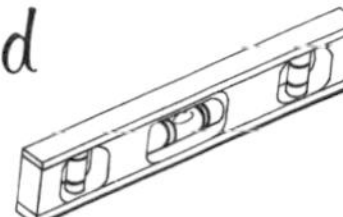

Adjustable wrench 

causing air to escape and the ice dispenser to leak. To level this appliance, you'll need to make adjustments to the refrigerator's feet, located in the front.

To find out if this appliance is level, check only the front part of the refrigerator, from right to left.

Place a carpenter's level on the top of the refrigerator, toward the front of the appliance. Look at the bubble in the level. For something to be level, the bubble must be centered between the two black lines. If the bubble slopes to the left, you'll need to *raise* the foot on that side or *lower* the foot on the right.

The feet are located at the bottom of the refrigerator and can be accessed by pulling off the lower panel. Using an adjustable wrench, turn the screw either clockwise to lower the feet, or counterclockwise to raise the feet.

Place the level on top of the refrigerator again, and if necessary, repeat the process.

Placing level on top of refrigerator

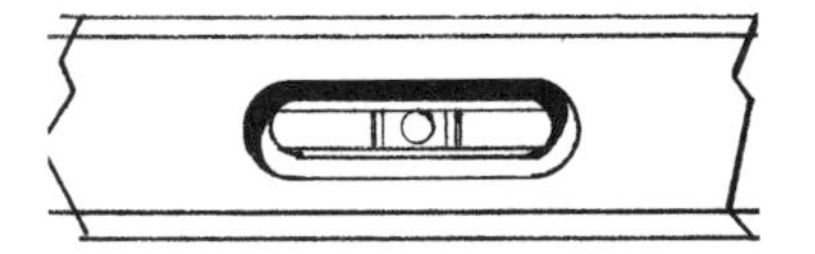

Checking position of bubble

Adjusting refrigerator's feet

Washing Machine

A washing machine that's not level can throw off the inner balancing ring and make the basket tilt, causing the appliance to stop. If the washing machine is very unbalanced, it can also cause the appliance to "walk," which means it actually moves. The majority of washing machines have self-adjusting back feet, so only the front feet need to be moved with an adjustable wrench. The back feet, as you'll soon find out, adjust themselves when you slightly lift up the back of the machine and drop it.

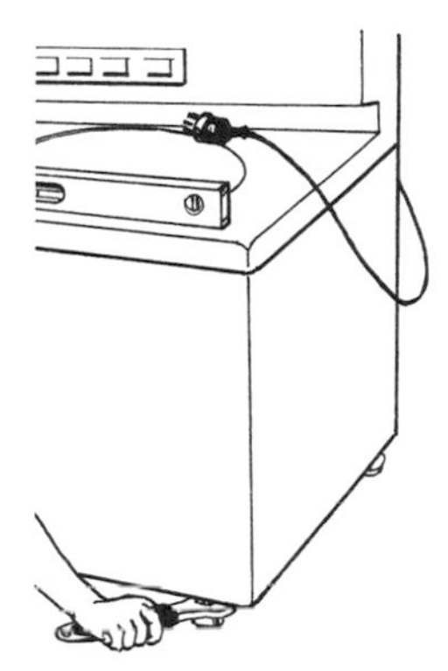

Adjusting washing machine's feet

To determine if a washing machine is level, you'll need to do two steps: 1) place the level from side to side on the top of the machine; and 2) place the level from back to front on the top of the machine.

Unplug the washing machine or, if you can't reach the outlet, turn *off* the power at the main service panel. Place the carpenter's level horizontally across the top of the washing machine, facing you. Look at the bubble in the level. For something to be level, the bubble must be centered between the two black lines.

The feet of the washing machine should be accessible without your removing any panels. Using an adjustable wrench, turn the screw clockwise to *lower* the front feet or counterclockwise to *raise* the front feet. Check the location of the bubble again. Repeat the process until the bubble is centered.

Next, place the level on top of the washing machine again, perpendicular to you. Check the bubble in the level. If the bubble is off center, you'll need to adjust the back feet. Take the level off the machine.

Tools Needed

Carpenter's level 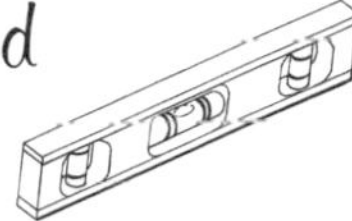

Adjustable wrench 

Using both hands, grab the back of the machine, pulling it toward you so that its back feet are about 4 inches off the floor. Let go of the appliance, allowing it to hit the floor (*it seems so naughty, doesn't it?*). Place the level on the washing machine again, perpendicular to you, and check the bubble. If the bubble is centered, use the adjustable wrench to tighten the screws on the front feet of the appliance. Repeat if necessary. Restore power to the appliance.

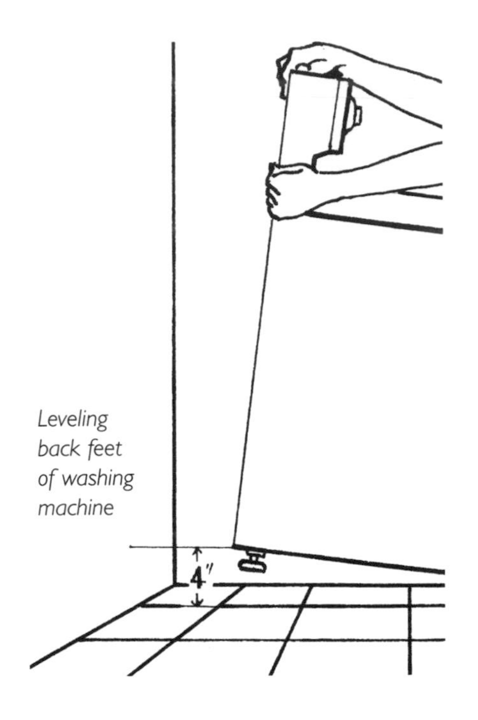

Leveling back feet of washing machine

Freeing a Jammed Garbage Disposal

A garbage disposal is not a necessity . . . until it's broken. We asked some women what the craziest things were that jammed their garbage disposals, and here's what they said: a baby food jar, a set of keys, and our personal favorite—a dozen red roses that our friend Pam decapitated one at a time (it was a liberating experience, she said).

If you hear an unusual humming or buzzing noise when you turn on the garbage disposal, it's jammed. Resist the urge to grind the object into oblivion—no matter how bad your day was! The garbage disposal's overload protector automatically shuts off the motor to protect it from burning out. This safety feature also serves to protect the electrical wiring within your home. Some disposals reset themselves automatically after 15 to 30 minutes. Other models have a reset button which, when pushed, resets it. This is found on the exterior base of the disposal motor.

Some garbage disposals have a reversal feature, which allows the flywheel (the rotating plate that shreds the waste) inside the disposal to

Never put chicken skin, grease, plastic, glass, meat bones, fruit peel and pits, or a drain cleaner into the garbage disposal.

Tools Needed

Flashlight 

Needlenose pliers or tongs 

¼-inch Allen wrench or broomstick handle 

operate in reverse to free the obstruction. The reversal feature is automatically activated each time the unit is switched on. Other garbage disposals have a hole in the exterior base of the unit where a ¼-inch Allen wrench can be inserted to turn the flywheel (some units come with the wrench). If in doubt, check the appliance owner's manual or contact the manufacturer.

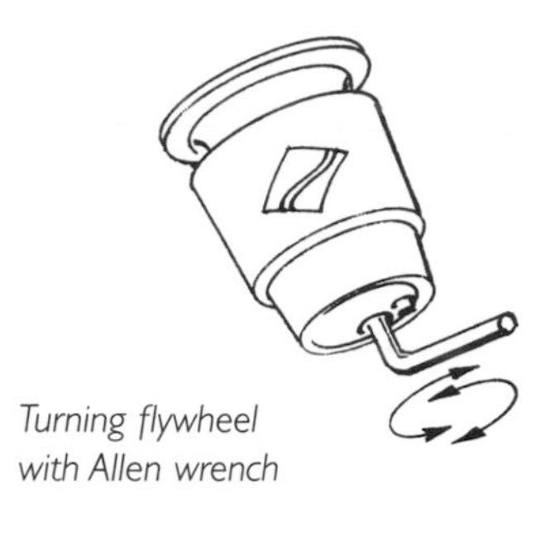

Turning flywheel with Allen wrench

First Try This

Turn the electrical on/off switch that operates the garbage disposal to the *off* position. Use the flashlight to look inside the disposal for the lodged object. Remove it using tongs or pliers.

If That Doesn't Work

If you can't locate the object *and* your disposal has a reversal feature, run the coldwater faucet and turn *on* the disposal. It may be necessary to turn the disposal *on* and *off* several times to activate the reversal feature.

If Your Disposal Does *Not* Have a Reversal Feature

Turn *off* the main power by removing a fuse or flipping the circuit breaker. Locate the hole on the bottom of the disposal under the sink and insert the Allen wrench into the hole. Work the wrench in both directions, counterclockwise and clockwise until it turns fully in complete circles.

If you don't have an Allen wrench, stick the broomstick handle into the disposal, resting it against the blades. Using the same technique as with the wrench, turn the handle counterclockwise and clockwise to dislodge the obstruction.

After freeing the jam, remove the dislodged material with tongs and restore the power. Push the reset button (which activates the overload protector) and run cold water for about 1 minute. Turn *on* the garbage disposal. If the disposal still won't start, wait 15 minutes for the motor to cool fully and press the reset button again.

If the object just won't budge, then contact the manufacturer for a list of professional repair persons in your area.

Allen wrench inserted in the bottom hole of garbage disposal

Finger depressing reset button on garbage disposal

Fixing a Dishwasher

It was Mother's Day and Maria Rosa's children had planned a special day for her. After serving her breakfast in bed, the kids loaded the dishes into the dishwasher along with *dish* detergent. Maria Rosa thought the laughter she heard was from her children delighting in cleaning the kitchen. Little did she know that the joyous sounds were from the kids playing in the massive amount of bubbles spewing from the dishwasher!

Solving Suds

It's easy to confuse dish and dishwasher detergents, because some containers look identical. If this mistake happens in your home, the following steps will easily solve the *Lawrence Welk* look.

Tools Needed

Plastic bowl 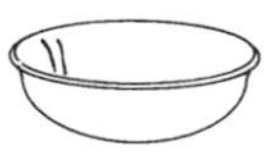

Measuring spoon (tablespoon) 

Shortening (plain, not butter-flavored; solid, not liquid)

Dishwasher detergent 

For homes with hard water, the most effective dishwasher detergent is a granular type; for soft water, use liquid or gel. If you don't know whether your water is hard or soft, call your water company or purchase a test kit from a hardware store.

Remove all of the dishes from the dishwasher. Use a plastic bowl to rid the dishwasher of as many of the suds as possible. Add 2 to 3 tablespoons of shortening to the bottom of the dishwasher and run a complete cycle. It sounds bizarre, but the shortening will cut the suds. Once the cycle is complete, add dishwasher detergent and run a second complete cycle. Add the dishes and dishwasher detergent, and run a final cycle. The bubbles should disappear.

When Dishes Aren't Getting Clean

If your dishes aren't getting completely clean, don't change detergents just yet. The problem could be that the dishes weren't properly cleared of food before being placed inside the dishwasher, or maybe the dishes were blocking the spray arm. The problem could also lie with a low setting on the water heater, which you can change yourself. Wait, there's more! It could also be that there's a faulty thermostat, a broken heating element (both of which you'd need a contractor to repair), or low water pressure, which would require the services of a certified plumber. *Phew.*

First Try This

Check the reading on the water heater, remembering that for safety reasons the temperature should be set at 120°F (see "Setting a Water Heater's Temperature" on page 220). Some dishwashers

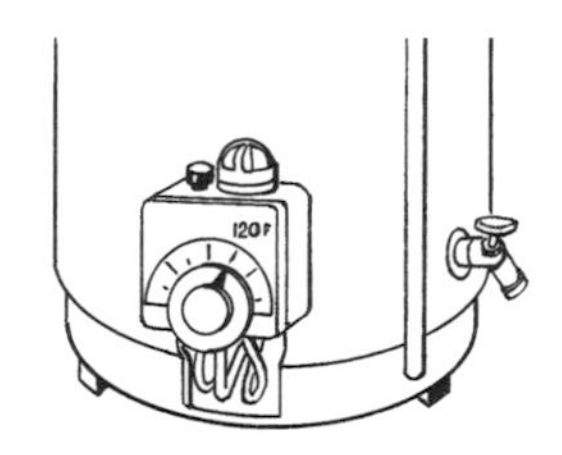

Temperature dial on water heater

Tools Needed

Owner's manual

Meat or candy thermometer

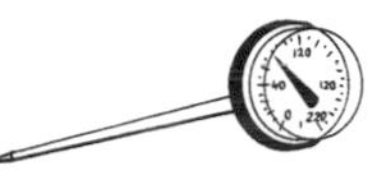

have a power boost that heats the water to a higher temperature than your water heater produces. Refer to the owner's manual or contact the manufacturer to find out if it contains this feature.

If That Doesn't Work

Turn the dishwasher *on* and open the door during the first cycle. Slide out the lower rack so that you can place a thermometer in the bottom of the dishwasher, which should be filled with hot water. Be careful to stay clear of the heating element—*muy caliente!* If the temperature reads less than 120°F *and* your water heater is correctly set, the problem could be the dishwasher's heating element or thermostat. Contact your appliance manufacturer for a list of professional repairpersons in your area.

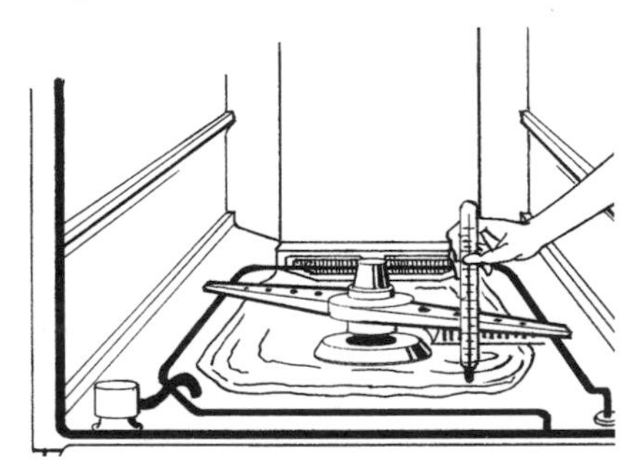

Testing the water's temperature

Maintaining a Refrigerator

Of all the appliances in your home, the refrigerator is the one you really can't do without. Clothes can be cleaned in a laundromat, dishes can be washed in the sink, and meals can be delivered. But not having a place to store the raw chocolate chip cookie dough would be a *travesty* (that is, of course, if you like that sort of thing). Therefore, we're providing you with two easy ways of helping to keep this appliance healthy for a long time.

Testing Refrigerator/Freezer Gaskets

We've all stood in front of an open refrigerator searching for that one special item that will satisfy our hunger. Meanwhile, cold air is escaping, causing the refrigerator to work a little harder to keep itself at the desired temperature.

A worn gasket will also allow cold air to escape from a refrigerator/freezer. A gasket is the rubber material located around the perimeters of refrigerator and freezer doors, which provides strong suction when the door is closed so cold air can't escape.

Tool Needed

Dollar bill

If a gasket cracks and loses its suction, the refrigerator/freezer will work overtime, causing your electric bill to increase and the life of the appliance to decrease.

Place a dollar bill in the door of the appliance and close the door. Gently tug the dollar bill. It will come out easily, but you should feel some resistance. If the dollar bill slips down, then the gasket isn't working properly. If the gasket is severely cracked and has lost all of its suction, you'll need to have it replaced by a professional repairperson, and ladies, it's not cheap!

Testing gasket with dollar bill

Maintaining a Gasket

Clean the gaskets with a soapy sponge or cloth and wipe completely dry. Using your finger, apply a small amount of petroleum jelly to the gaskets to prevent cracking.

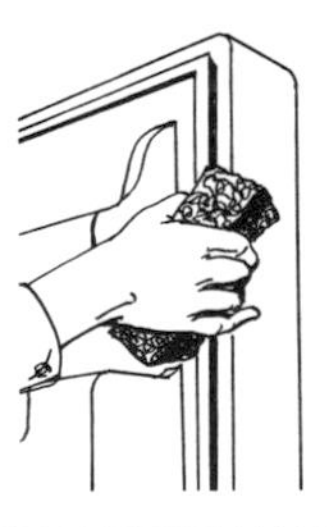

Cleaning gasket

Applying petroleum jelly to gasket

Tools Needed

Sponge in mild soapy water  **Towel**  **Petroleum jelly** 

Vacuuming Refrigerator Condenser Coils

Is your refrigerator running? Well, you better go catch it! Brings back memories of sleepover parties *before* Caller ID. But do you really know what to do if your refrigerator is running? A running refrigerator is one that is overworking itself because its coils are dirty, it has a faulty gasket, or there are problems with its condenser fan or motor.

Refrigerators are quiet appliances, so you can tell the appliance is struggling if it's emitting a loud humming or buzzing noise. Sometimes you can become so used to a loud refrigerator that the sound becomes white noise. Even if the noise isn't bothering you, the cost of replacing the refrigerator will resound loudly. It's also important to solve the problem quickly to keep your electricity bill down.

Refrigerators have condenser coils located on the back or on the bottom, behind a grille. The coils should be cleaned twice a year, or more if you have pets.

If your refrigerator has its condenser coils on the back, move the refrigerator out from the wall with your friend's help. Be careful not to tip the refrigerator while moving it, because it could fall over. Unplug the refrigerator. Vacuum the condenser coils using

Moving refrigerator

Tools Needed

Helpful friend

Flashlight

Vacuum cleaner (with brush or wand attachment)

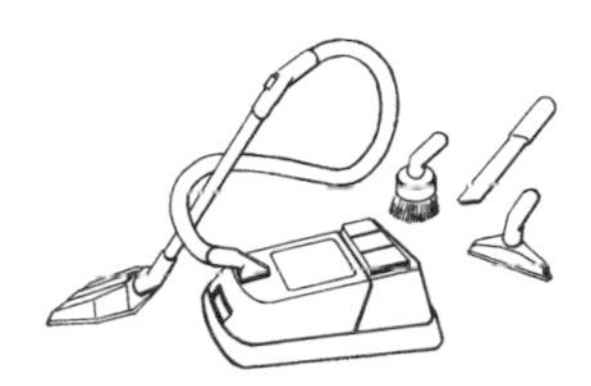

the brush attachment. Plug the refrigerator in, and carefully push it back into place.

To clean condenser coils located underneath the refrigerator, move the refrigerator (with your helpful friend, if necessary), but only far enough to unplug it. Remove the grille, located at the base of the refrigerator. Holding the flashlight with one hand, clean the condenser coils using the vacuum with the wand attachment. Replace the grille. Plug the refrigerator in, and carefully push it back into place.

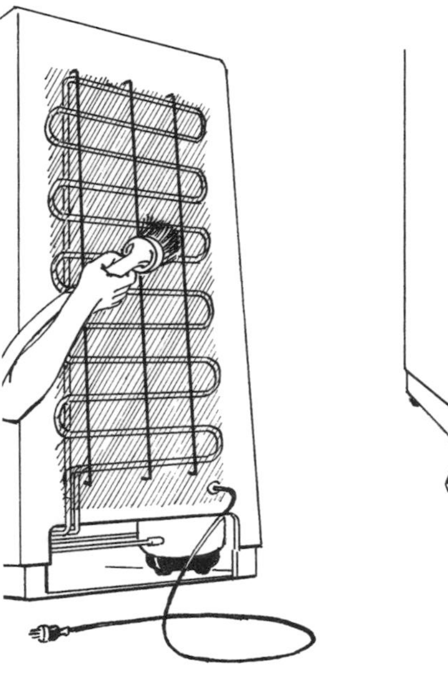

Vacuuming condenser coils on back of refrigerator

Vacuuming condenser coils underneath refrigerator

Cleaning a Stove's Grease Filter

True or False: *A grease filter is located under the hood of your car.* You need to use a lifeline, don't you? The answer is *False.* A grease filter is located in the exhaust system of a range (including island cooktops and hood-mounted microwaves).

A grease filter, which is made of metal, protects the motor fan by trapping dirt, dust, and grease in the air that's being pulled through the exhaust system to the outdoors. A charcoal filter is used in conjunction with a metal filter for exhaust systems that recirculate the air internally (e.g., hood-mounted microwaves). It's located behind the metal filter and eliminates odors in the air.

Stovetop and hood with exposed grease filters

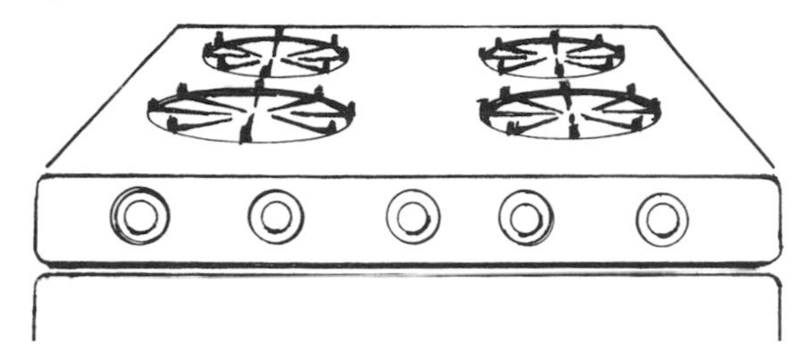

A metal grease filter should be cleaned regularly and replaced every 3 to 4 years, or when it begins to fray. A charcoal filter can't be cleaned and should be replaced every six months. Filters can be purchased at appliance and hardware stores, or through the manufacturer.

Cleaning a Metal Grease Filter

Remove the filter and clean it in a dishwasher. If this is the first time you've cleaned the filter, you may want to soak it first before putting it into the dishwasher. If you have to wash the filter in a sink, use hot soapy water. While the filter is out, clean the area where the filter is housed with a soapy sponge or cloth. Completely dry the filter before returning.

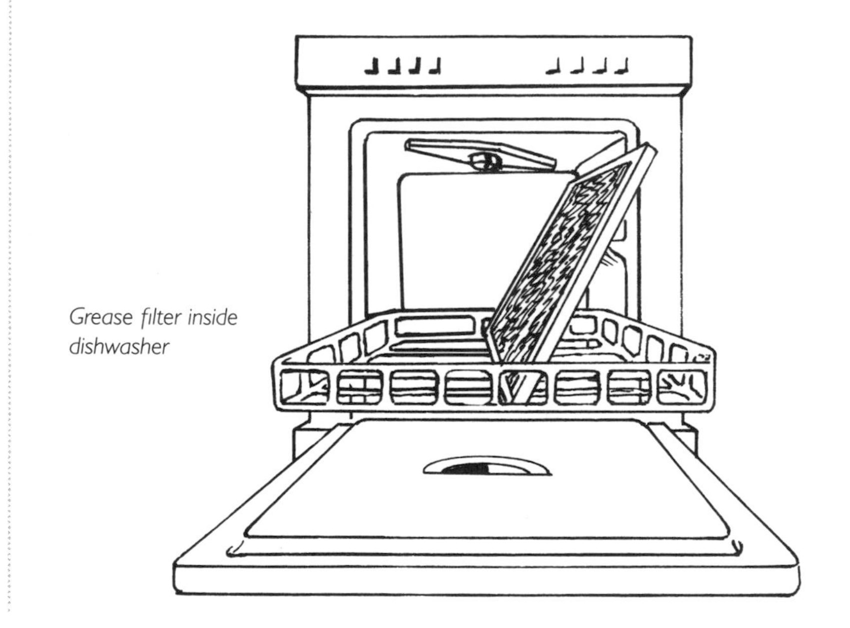

Grease filter inside dishwasher

Tool Needed

Sponge or cloth in hot soapy water

Repairing a Slow-Filling Washing Machine

The Novellas is a women's book club bound together by years of meetings and shared lives. But after fifteen years, the members decided they needed to add a new chapter to their lives by becoming a home repair club. Each month a member chose a repair that needed to be done in her home and everyone arrived ready for the challenge. The first repair was fixing a slow-filling washing machine. It didn't take long for those female bookworms to be repair-literate, too, and the repairs were so easy no one needed Cliff Notes!

If your washing machine is taking a long time to fill up with water, the likely culprits are the inlet screens. These are located inside the inlet valve holes, which are found on the back of the washing machine where the hot and cold hoses connect to it. These small screens protect the hoses from becoming clogged with debris. If particles are trapped inside the hoses, water is not able to flow freely into the washing machine.

If you can't find screens in your washing machine inlet valves, there could be two reasons: 1) in areas where hard water is prevalent, inlet filters are deliberately not installed because the sediment in hard water can clog a screen very quickly and most people won't go to the trouble of cleaning them; or 2) the screen is strategically placed in a hard-to-get-to spot so that people won't remove it. When in doubt, refer to the owner's manual or contact the manufacturer.

Unplug the washing machine, or turn *off* the power at the main service panel, and pull it out as far as possible so you can get behind it. Turn *off* the water supply valve(s) located on the wall behind the washing machine.

You'll see that the washing machine has two inlet hoses—one for hot water and one for cold (hot is typically located on the left; cold on the right). Each hose is attached to the appliance at the inlet valves, and each valve contains a small inlet screen.

Before removing the hoses, tear off two pieces of masking tape and mark one "hot" and the other "cold," and tape them to their respective hoses. Use the slip-joint pliers to loosen the coupling nuts by turning them counterclockwise. Remove the hoses and pour the water into the utility sink or bucket.

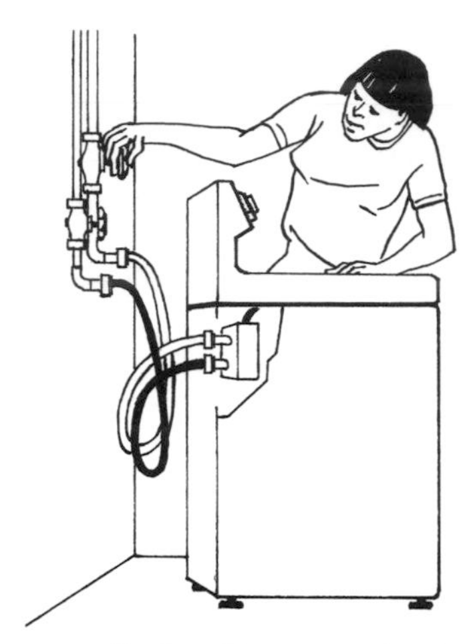

Turning off water supply valve

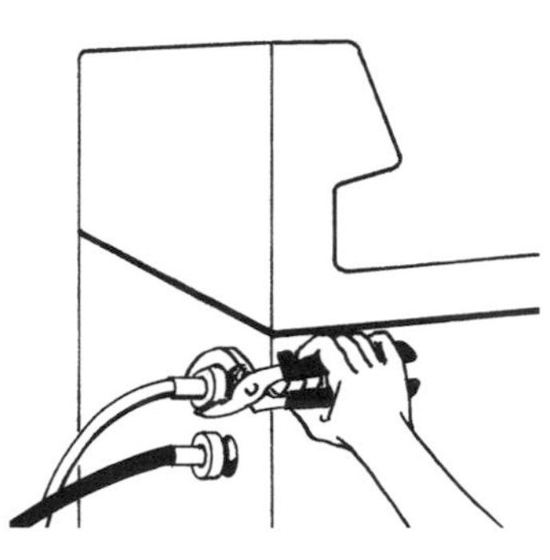

Loosening coupling nuts

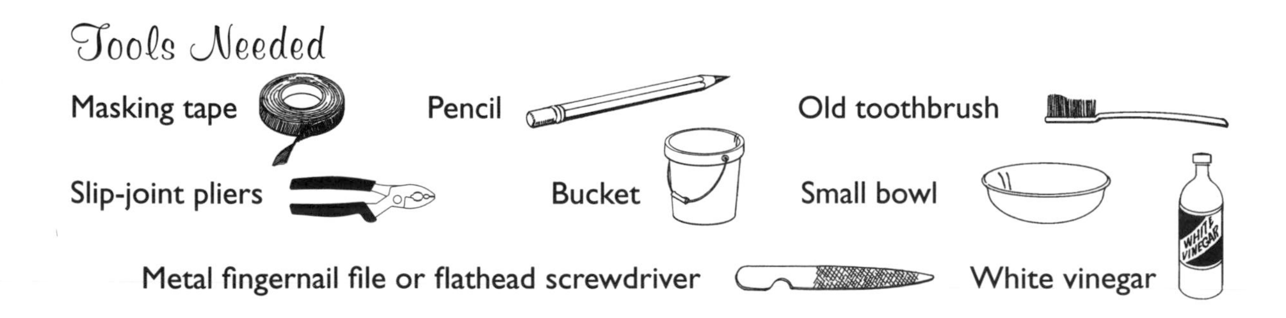

Locate the inlet screens in the valves and pop them out with a metal fingernail file, being careful not to tear the screens. You can choose to clean the inlet screens or replace them with new ones. To clean the screens, soak them in white vinegar for 15 minutes, gently scrub with the toothbrush, and rinse with warm water. Return the old or new inlet screen to each valve.

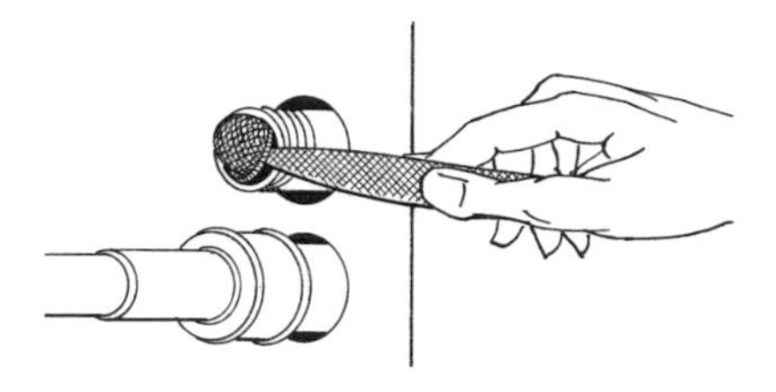

Removing inlet screens

Reconnect the hoses to the washing machine, noting the labels on the tape. Tighten the coupling nuts by turning them clockwise, making certain not to overtighten. Restore power to the appliance, turn *on* the water, and push the washing machine back into place.

Cleaning a Clothes Dryer Hose and Vents

Kim was thrilled to finally have someone fix the clothes dryer that had been broken for weeks. The repairman arrived at her door with tools on one side and a helper on the other. As Kim watched one man clean the vents and hose, the able-bodied assistant was busy cleaning out her wallet. Kim's dryer got fixed, but she got taken to the cleaners.

Did you know that over 90 percent of the complaints an appliance manufacturer's consumer hotline receives about dryers is that the clothes are taking too long to dry?

If you're running two cycles to get a load of clothes dry, there's probably a clog in the appliance's hose and vents caused by a buildup of lint. Another way to tell if there is a clog is to put your hand on the exterior exhaust vent (outside your home) and feel for a healthy flow of hot, wet air while the dryer is running. If you don't feel anything, then a clog exists.

A dryer's hose can become a fire hazard if it becomes clogged with lint and/or it's not the proper type of hose. An electric and gas

Tools Needed

Vacuum with long attachment, or broomstick with a towel (secured with rubber band)

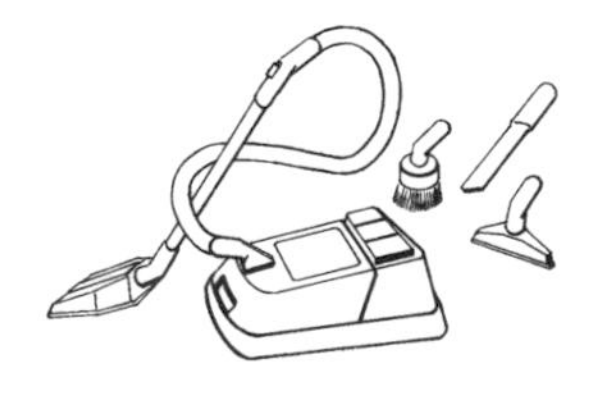

Old toothbrush

dryer's hose should be constructed of metal only (rigid or flexible) because if a fire starts, the hose will contain it inside the dryer. You should avoid hoses made of plastic or a slinky type of aluminum foil. If you're not sure which type of hose your dryer has, contact the manufacturer.

We recommend you do this project yourself only if the dryer is located very close to the exterior exhaust vent. The farther your dryer is from the vent, the longer the hose, which makes it more susceptible to clogging. Therefore, you should have the hose professionally serviced every year, or according to the manufacturer's guidelines.

Turning Off the Gas

Unplug the dryer. If it's a gas dryer, turn *off* the gas supply. (The gas handle is usually located on the gas line where it connects to the dryer.) When the handle is in line with the pipe, it means that the gas is *on*. To turn the gas *off*, move the handle 90 degrees. If you have trouble with the handle, contact your gas company for a list of certified plumbers in your area.

Turning off gas supply to dryer

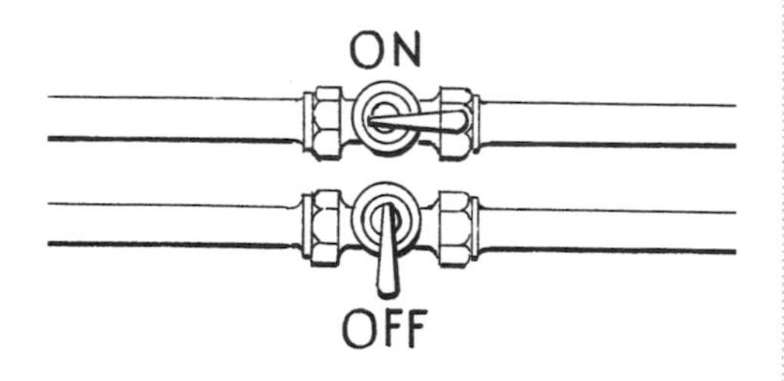

Identifying "on" and "off" positions

Owners of gas dryers should be careful not to let clothes on the floor get too close to the dryer, because the pilot is located at the bottom and the clothes can easily catch fire.

Cleaning the Vents and Hose

Pull the dryer out so you can get behind it. Rotate the hose where it connects to the back of the dryer (you may need to loosen it with a screwdriver) and at the wall exhaust vent and remove. Lay the hose on the floor.

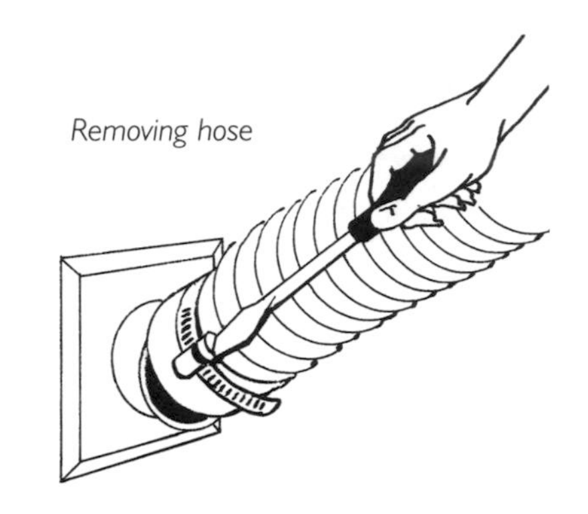

Removing hose

Plug in the vacuum cleaner (with the long attachment), and place it in one end of the hose to remove the lint. If you don't have a vacuum with a long attachment, use a broom handle (with the towel) and push it in and out of the hose until it's clean. Use the toothbrush to remove any stubborn lint found in the exhaust hood and dryer outlet. Reattach the hose to the dryer outlet and the exhaust vent. Turn *on* the gas supply, if applicable. Plug in the dryer.

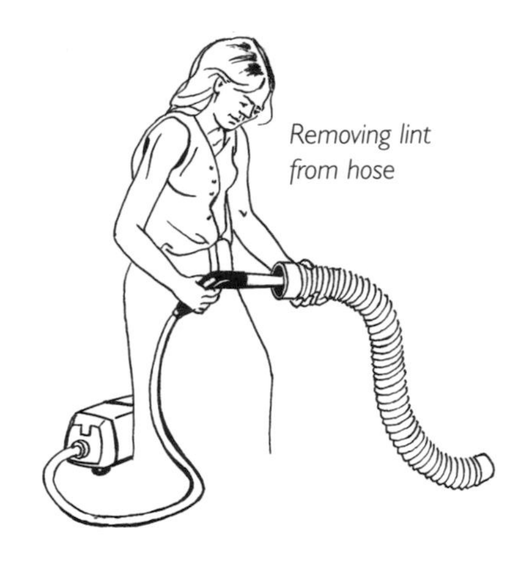

Removing lint from hose

Cleaning or Replacing a Furnace Filter

When Gretchen inherited her great-aunt's house, she never thought to have it inspected, because the house was free—why waste the money? If she had, she would have known, that the furnace was about to bite the dust. Actually, the furnace had been choking on dust for years because the filter had never been cleaned or replaced. If only Gretchen *had* looked the gift horse in the mouth!

Basic maintenance for a heating/cooling system is easy—clean the air filter every month and have it professionally serviced once a year. An air filter's job is to trap the dirt and dust particles before they enter the furnace system. By keeping the air filter clean you allow the heating/cooling system to work efficiently, which means that you keep your energy bills down while prolonging the life of your system.

If you've seen the letters HVAC and wondered what they mean, it's an acronym for Heating, Ventilation, and Air Conditioning.

Purchasing a Filter

When purchasing a filter, it's important to know that all filters are not the same size *and* you get what you pay for. Take your filter to the hardware store so you can be assured of getting the right size. And always check the manufacturer's instructions and warranty before purchasing. If you see a filter that claims to be "permanent," don't be fooled into thinking that it never has to be replaced—the lifetime warranty is for the frame only, not the filter screen.

Emergency power switch and thermostat

If you have a gas furnace, locate its emergency power switch. The switch plate should be red and say "Emergency Switch," but if the contractor who installed it didn't want to spend the extra buck for the special switch plate, then it can look just like any other light switch. The emergency switch can be found either on the furnace or in the same room—or very near it—where the furnace is located, depending on who installed it. Turn the switch *off*. If you have an electric furnace, turn it *off* at the thermostat.

Removing the Old Filter

Some filters are located in a slot between the furnace and large return air duct and can be removed by pulling out the slot where the filter is

Some furnaces may have two slots for filters; you will need a filter for each slot.

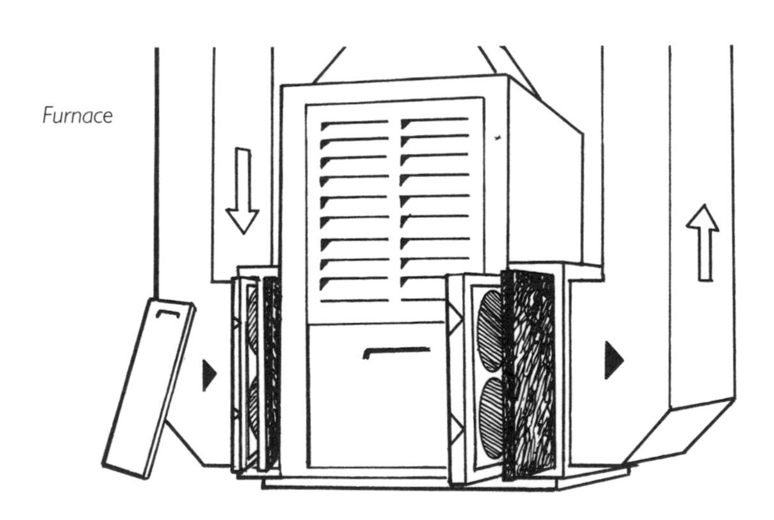

Tools Needed

Garden hose or bathtub/utility sink 

Furnace filters 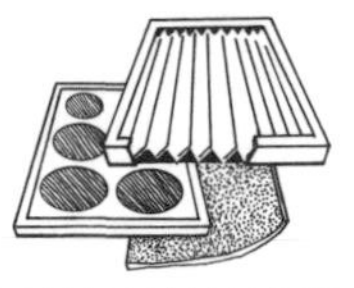

housed. Filters can also be found behind the front panel of the furnace. Once the panel is removed you'll find a large metal clip, which keeps the filter in place. This clip has to be snapped out before the filter can be removed.

Cheap filters come in sheets that can be cut to the size you need. If you're paying a dollar or less for a filter, you're not getting the top of the line model and it will have to be thrown out every month. A better way to save money is to buy one that can be cleaned and reused.

Cleaning the Filter

If the filter is a disposable model, throw it out. If it isn't, take the filter outside and clean it with a garden hose, front and back. Or, place the filter in a bathtub or utility sink and run water through it. Hold it up to the light to see if you've missed any spots.

Let the filter dry completely before returning it to the furnace. Place the new or cleaned filter in the slot, or put it behind the clip and replace the front panel. It's important to refer to the arrow on the filter when returning it to the furnace—the arrow should point away from the return-air duct and toward the blower. If the filter is put in backward, it won't work efficiently. Restore the power to the furnace by flipping the switch *on*.

Cleaning furnace filter

Lighting a Pilot in a Gas Furnace

It was a bitter cold December evening, and Mary had a house full of overnight guests visiting from Mississippi. Little did she know she'd wake up the next morning to a freezing cold house and guests who wished they'd never crossed the Mason-Dixon line. It seems that one of the children had accidentally flipped the furnace's emergency switch *off* the night before. Because Mary didn't know the furnace had that switch, she ended up paying a technician fifty dollars just to flip it *on*!

Ignorance isn't bliss . . . it's costly! It pays to become familiar with your home's mechanical systems, especially *before* an emergency happens.

A gas furnace typically lasts between 17 and 23 years, but there are a lot of other reasons besides age (and the emergency switch being flipped off) that can cause it to stop working—the gas to your home was shut off, a circuit breaker was tripped or a fuse was blown, filters were not cleaned or changed, the pilot light went out, or a number of internal problems occurred.

If your furnace and water heater are located in an open area where children play, use tape, string, chalk or paint to establish a line of demarcation around the appliances and tell the children never to cross over the line.

Do You Have a Manual or Electric Ignition?

There are two types of gas furnaces: those with a standing pilot and those with an electronic ignition. A standing pilot, which when operating correctly produces a blue flame, ignites the burner located at the base of the furnace. When a standing pilot goes out, it has to be manually lit.

An electronic ignition model lights the burner electrically and is typically found in newer gas furnaces. If the burner goes out, you can't light it yourself. Before calling a HVAC contractor, check another gas appliance or contact the gas company to see if the gas supply was shut off.

How can you tell whether your gas furnace has a standing pilot or an electronic ignition if the flame is out? An electronic model has a cord that plugs into a 120-volt outlet or other power source.

Safety Rules Before Beginning

First, sniff the area for gas. Natural gas is colorless and odorless, but the gas company adds a strong odor to it so you can detect a gas leak. If you smell gas, close the gas shut-off valve to the furnace. Leave your home immediately and contact the gas company from a neighbor's home or your cell phone. If you can't reach the gas company, call the fire department.

It's also important to read the instructions on the furnace and/or in the owner's manual before beginning so you can familiarize yourself with the process and the parts. If you don't have either, call the manufacturer.

Turn the thermostat to the lowest possible setting. Turn *off* the electrical power at the main service panel or flip the electrical emergency switch *off*. (The switch plate should be red and say "Emergency Switch." However, the location and color of the switch often depends upon the contractor who installed it.) The switch can be found on the

When the valve is in line with the gas line, the pipe is open; if it's perpendicular to the gas line, it's closed.

Tools Needed

Owner's manual

Screwdriver (Phillips or flathead) 

Watch 

Barbecue lighter or long match 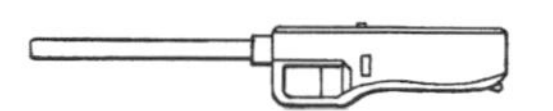

furnace, on the wall near the furnace, or sometimes alongside the room's main light switch. No matter where the switch is located, it's wise to have the red "Emergency Switch" cover to prevent accidents. The switch cover can be purchased at a hardware store for about a dollar.

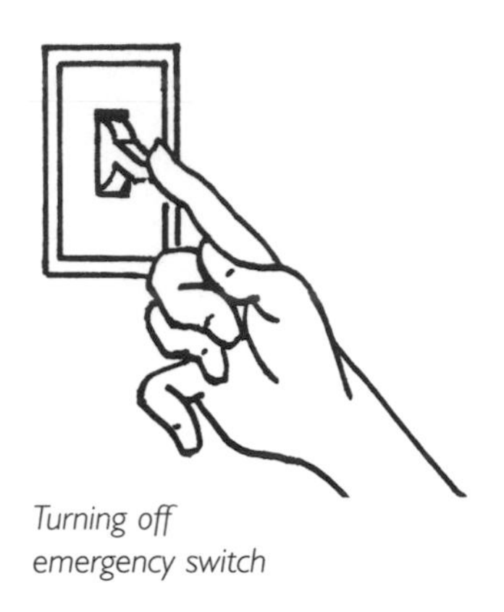

Turning off emergency switch

Check that the gas control knob (a.k.a. gas cock knob) is turned to the *off* position. Wait at least 10 minutes to allow airing out of any gas fumes that have accumulated in the burner chamber. If you smell gas, close the gas shut-off valve to the furnace. Do *not* turn on any lights or use your phone, because an electric spark can ignite the gas. Instead, leave your home

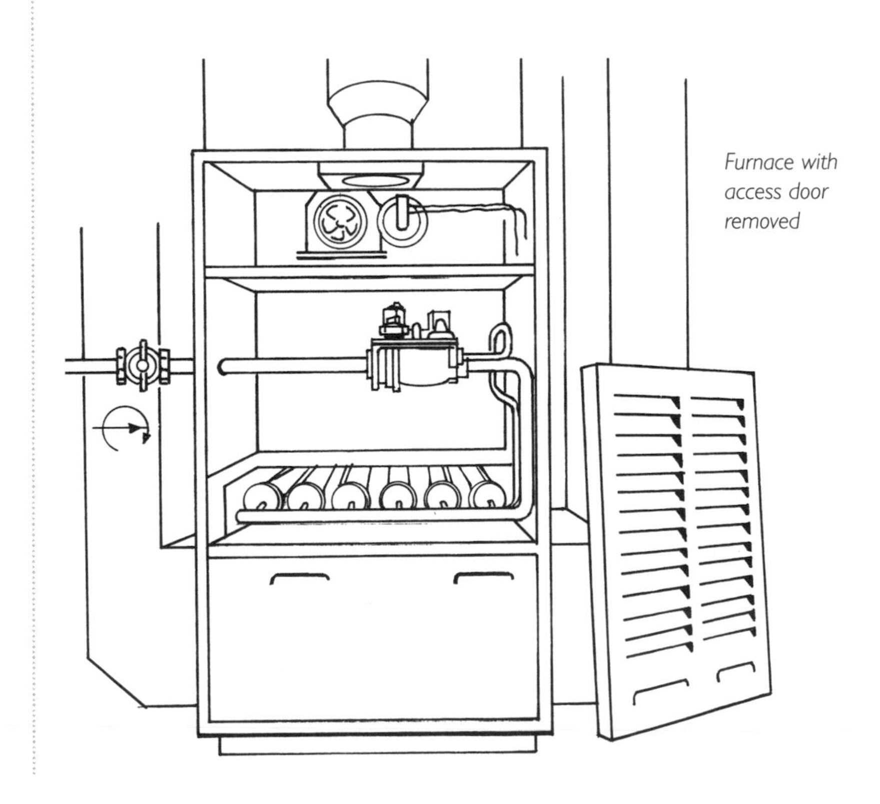

Furnace with access door removed

immediately and contact the gas company from a neighbor's home or use your cell phone. If you can't reach the gas company, call the fire department.

If your furnace has an access door to the burner chamber, remove the screws (if applicable) and lift or slide the door off.

Turn the gas control knob to the *pilot* position. Press down on the reset button (typically red), located next to the gas control knob. If your gas furnace does not have a reset button, depress the gas control knob. While holding it down, light the burner with the barbecue lighter, keeping your face turned away for protection. Continue to hold down the reset button for 1 minute after the pilot is lit and remove the lighter. Take your finger off the reset button and let it pop back up.

Igniting furnace burner without a reset button

If the standing pilot remains lit, replace the access door, if applicable. Turn *on* the emergency switch and set the thermostat to the highest position until the burner ignites. Adjust the thermostat to the desired temperature.

Windows, Walls, and Doors

- Disengaging an Electric Garage Door
- Removing a Broken Key from a Door Lock
- Patching Small Cracks and Holes in a Wall
- Freeing a Stuck Window
- Replacing a Doorknob
- Adjusting a Screen/Storm Door Closer
- Maintaining Window and Door Screens

This is the potluck section of the book—it has a little bit of everything, but not too much of any one thing. There's a dollop of doors, a pinch of windows, and a schmear of walls. No matter the subject, the information is designed to be easy to understand and complete so you can add another notch to that tool belt of yours!

Disengaging an Electric Garage Door

Alexia was an information junkie and prided herself on being the *go-to* person in the office. So, the morning she tried to leave home and the garage door wouldn't open, she was dumbfounded that she didn't know how to fix it. Alexia quickly called her assistant, Erin, for help. Seizing the moment, Erin replied, "Just how badly do you want to get out?" After bargaining for release in return for a vacation day, Erin told her how to disengage the garage door. "You didn't tell me it was that easy!" Alexia yelled. "You didn't ask," retorted Erin. Knowledge *is* power.

Don't panic and make an emergency phone call just yet. By following our instructions along with the information in the owner's manual, you're bound to be off and running in no time.

First Try This

If you've tried the remote control and it doesn't work, get out of your car and press the door switch located on the wall. If that works, you may only need to change the battery in the remote control.

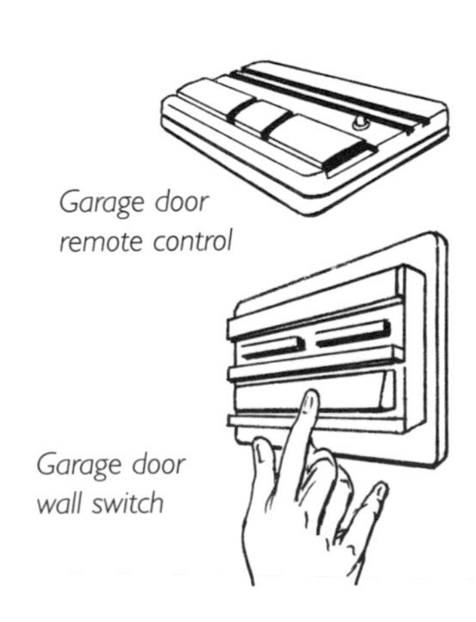

If That Doesn't Work

If the door switch doesn't work, there could be a number of reasons, such as the door was manually locked, there's an accumulation of snow or ice outside the door, or the circuit breaker was tripped or a fuse blew, causing the electricity to turn off. If after you've checked all of those things the door still won't budge, refer to your owner's manual and disengage it.

Releasing the Electric Garage Door

You probably know by now that the garage has great acoustics for screaming! Using the owner's manual, locate the manual release handle, which hangs from a rope connected to the trolley (a.k.a. carriage). Pull down hard. You'll know you did it correctly when you hear the loud sound of the opener disconnecting. Remember to manually lock the garage door once you leave.

Disengaging an electric garage door

Tool Needed

Owner's manual

Removing a Broken Key from a Door Lock

When Joanna broke her key in the door lock, she knew from experience that the super would charge a lot of money to open it, so she went to her elderly neighbor, Leona, who had her spare key. The independent octogenarian grabbed a pair of pliers, walked across the hall, and removed the broken key. Leona handed Joanna the spare key, along with some words of advice: "Honey, when life hands you a lemon, don't let it sour your day. Just take out your toolbox and fix it!"

Your ability to remove a broken key from a door lock depends on the degree of the break—if the key is visible, you can probably remove it yourself. But if you have only one entry door *and* you don't see the key, you'll have to call a locksmith.

To remove a broken key that is visible, lightly apply the lubricating spray to the key and lock. Using the needlenose pliers, grip the key and pull it out. If the key is broken off too far inside the lock, you'll need to call a locksmith.

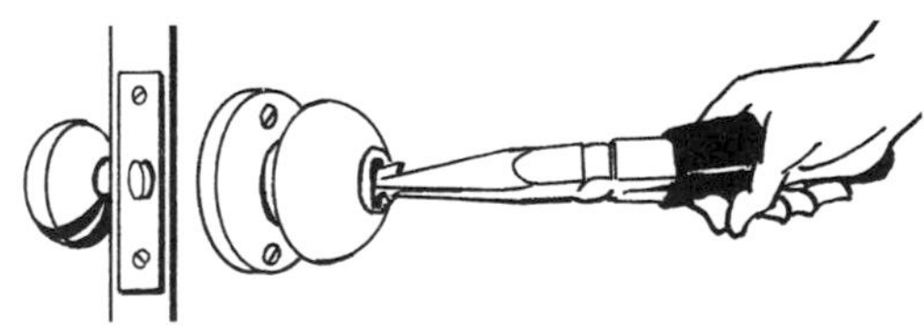

Removing a broken key with pliers

Tools Needed

Lubricating spray

Needlenose pliers or tweezers

Patching Small Cracks and Holes in a Wall

If you've ever iced a cake (*Easy Bake Oven* cakes do count), this repair will be a piece of . . . well, let's just say it's easy.

A crack in a wall won't cause you seven years of bad luck. It can, however, increase in size over time, causing you more work.

There are different types of drywall compound (a.k.a. mud) for different types of jobs. Use a lightweight compound rather than joint or spackling compounds for repairing cracks and holes because it's easier to apply and sand, it dries faster, and it comes in small containers.

Repairing a small crack or hole in a wall is an easy job, but it takes a while to complete because of the drying time. To speed things up, set a fan in front of the mudded area or use a hair dryer on a cool setting.

Repairing Cracks

Place newspaper on the floor under the area where you'll be working and cover any furniture nearby to protect it from drywall and dust.

Tools Needed

Newspaper 
Old credit card 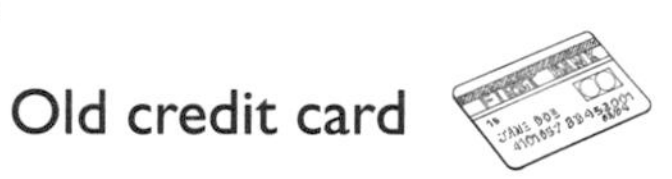
Utility knife 
Scissors 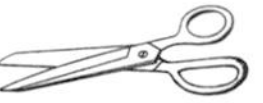
Old toothbrush 
Wall repair tape 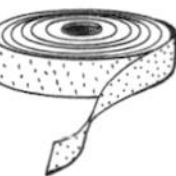
Putty knives (2" and 4") 
Wallboard sanding block (80 coarse grit, 150 fine grit) 
Lightweight drywall compound (a.k.a. mud) 

Mud tray 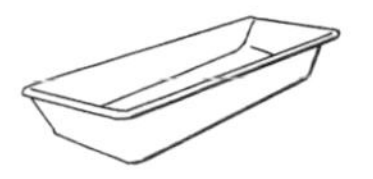

A crack can't be fixed unless it's wide enough to be filled with compound. The way to test the width is to place an old credit card in the top of the crack and run it all the way down. Wherever the card gets stuck is where you'll need to widen the crack.

Preparing the Surface

Use the utility knife to widen areas and to cut any rough edges around the crack, being careful not to get carried away with the cutting or you'll end up with a bigger job. If necessary, sand the area gently with the 150 grit (if you sand too hard, you'll create more work for yourself). Sweep away debris inside the crack with the toothbrush. Be sure to remove any dust; otherwise, the tape won't stick.

Taping the Crack

Measure the wall repair tape against the crack to see how much you'll need, adding a little extra to both ends. Cut the tape with the scissors and stick it onto the crack with the adhesive side down. Smooth it with your hand, top to bottom.

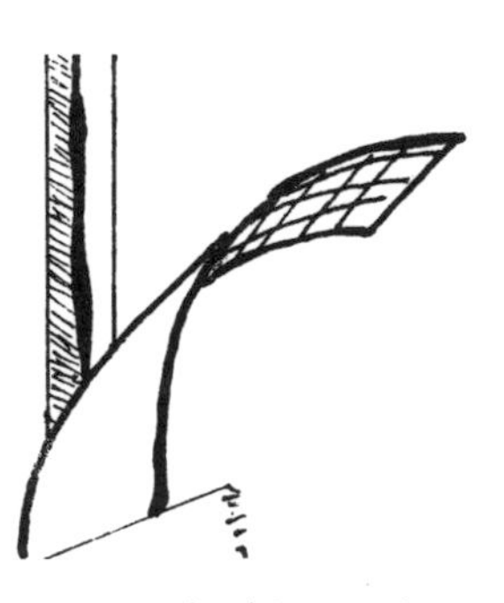

Applying repair tape

Applying the Compound

Use the 2-inch putty knife to add the compound to the mud tray and to apply the first coat to the entire taped area. The first coat is referred to as the "working" coat because it doesn't matter how it looks—you just need to apply a generous amount to the area. Scrape off any extra mud onto the edge of the tray. Clean the putty knife and tray, and wipe dry.

Allow the area to completely dry, typically about 8 hours. Using the coarse side (80 grit) of the sanding block, sand over the area to smooth out any wrinkles or bumps.

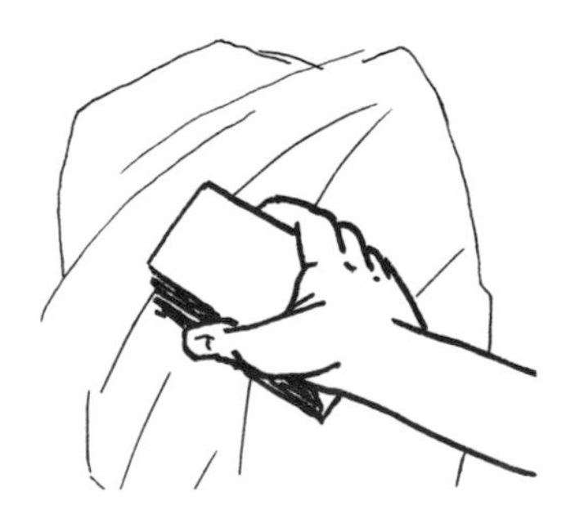

Sanding area

Adding Additional Coats

Use the 4-inch putty knife to add the compound to the mud tray and apply the second coat to the entire area. The second coat is referred to as the "cosmetic coat" because this time appearance does count; therefore, you need to apply a thin coat to the area. Clean the putty knife and tray, and wipe dry.

After the area is completely dry, use the fine side (150 grit) of the sanding block to smooth out any additional wrinkles and bumps. If necessary, apply another thin coat using the 4-inch putty knife. Let dry and sand again.

Once the tape and mudding blemishes are no longer visible *and* the area is completely dry, you can paint.

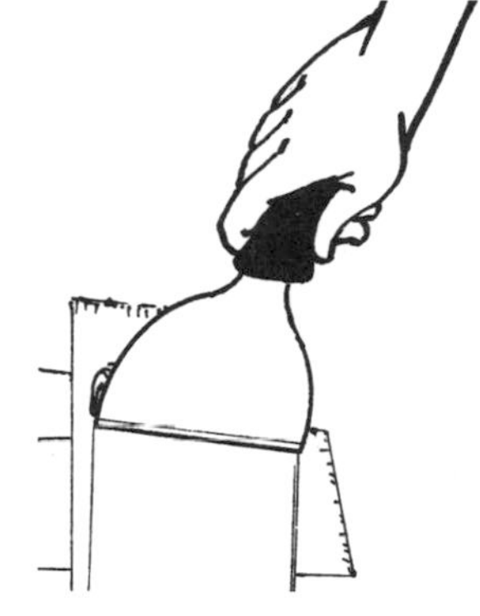

Applying mud

Repairing Holes

This repair is for patching holes no bigger than 4 to 5 inches in diameter (a larger hole may require installing a piece of drywall). For repairing nail

Tools Needed

Newspaper 

Utility knife 

Wall repair patch 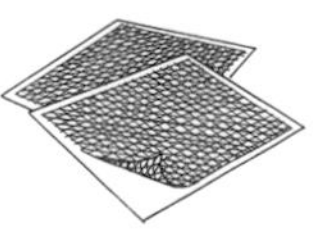

Wallboard sanding block (80 coarse grit, 150 fine grit) 

Scissors 

Lightweight drywall compound (a.k.a. mud) 

Putty knives (3" and 6") 

Mud tray 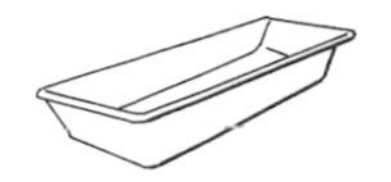

holes, apply a small amount of mud to your index finger and dab it onto the hole. Wipe away any excess mud and if necessary gently sand.

Place newspaper on the floor under the area where you'll be working and cover any furniture nearby to protect from drywall and dust.

Preparing the Surface

Use the utility knife to cut any rough edges around the hole, being careful not to get carried away with the cutting or you'll end up with a bigger job. If necessary, sand gently and wipe away any dust.

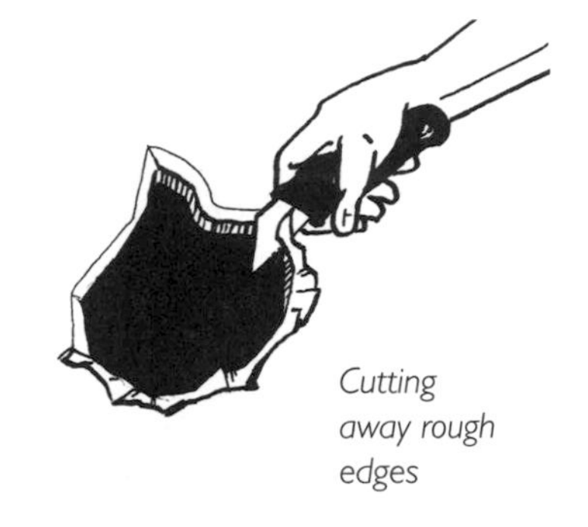

Cutting away rough edges

Applying the Repair Patch

Measure the wall repair patch against the hole to see how much you'll need, adding a little extra to all four sides. Cut the patch with the scissors and stick it on over the hole with the adhesive side down. Smooth it with your hand, top to bottom.

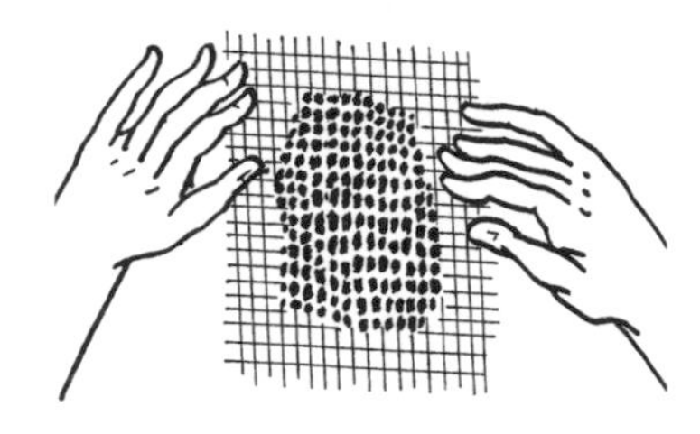

Measuring wall patch

Applying Compound

Use the 3-inch putty knife to add the compound to the mud tray. Apply a generous first coat with the putty knife. Starting on the wall above the patch, spread the mud up and down, side to side, and corner to corner. Scrape off any extra mud onto the edge of the tray. Clean the putty knife and tray and wipe dry.

Applying first coat of mud

Allow the wall to completely dry, typically about 8 hours. Using the coarse side (80 grit) of the sanding block, sand over the area to smooth out any wrinkles or bumps.

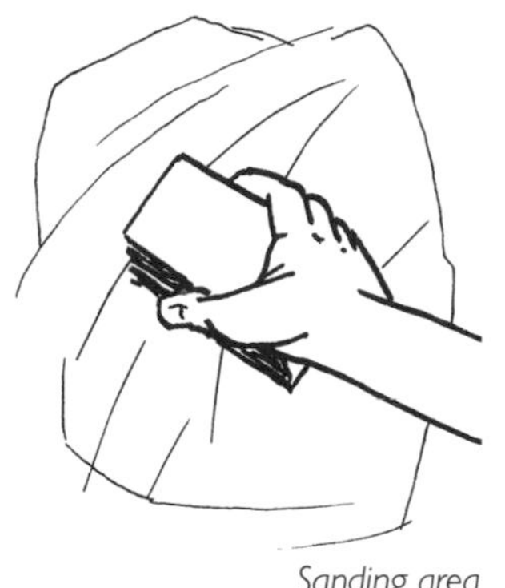

Sanding area

Adding Additional Coats

Add compound to the mud tray with the 6-inch putty knife and apply the second coat to the entire area, using the same technique as above. Clean the putty knife and tray and wipe dry.

After the area is completely dry, use the fine side (150 grit) of the sanding block to smooth out any additional wrinkles and bumps. For the final coat, use the 6-inch putty knife to apply a thin layer of mud twice the size of the hole, again spreading up and down, side to side, and corner to corner. Let dry, and sand using the fine side of the block.

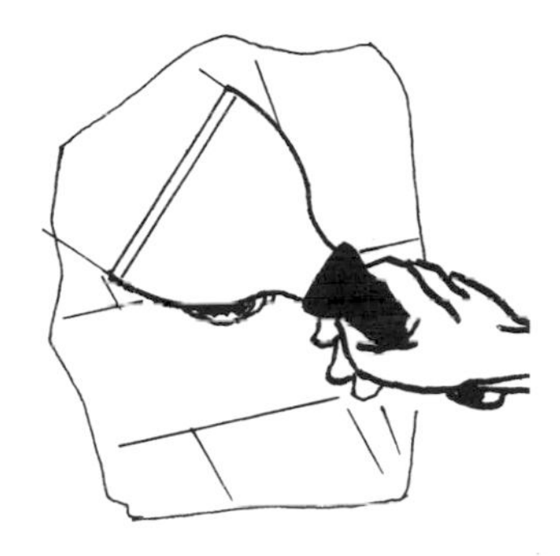

Applying second coat of mud

Once the hole and mudding blemishes are no longer visible and the area is completely dry, you can paint.

Freeing a Stuck Window

Frank Sinatra wasn't the only one who liked the *Summer Winds!* Don't let a stuck window deprive your home of some good fresh air. Gather up the tools needed and get started on this project now. It's really a breeze!

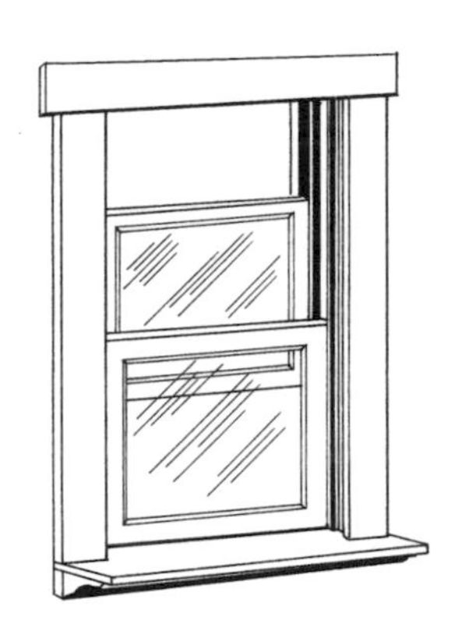

If you can't open your window, it's stuck because of humidity or dirt buildup, or it may have been painted shut. Of course, there's also the chance that the owner or tenant before you was very security conscious and may have nailed it closed. If so, remove the nails with needlenose pliers, or wrap masking tape around the head of the hammer (to prevent marring) and pull the nails out.

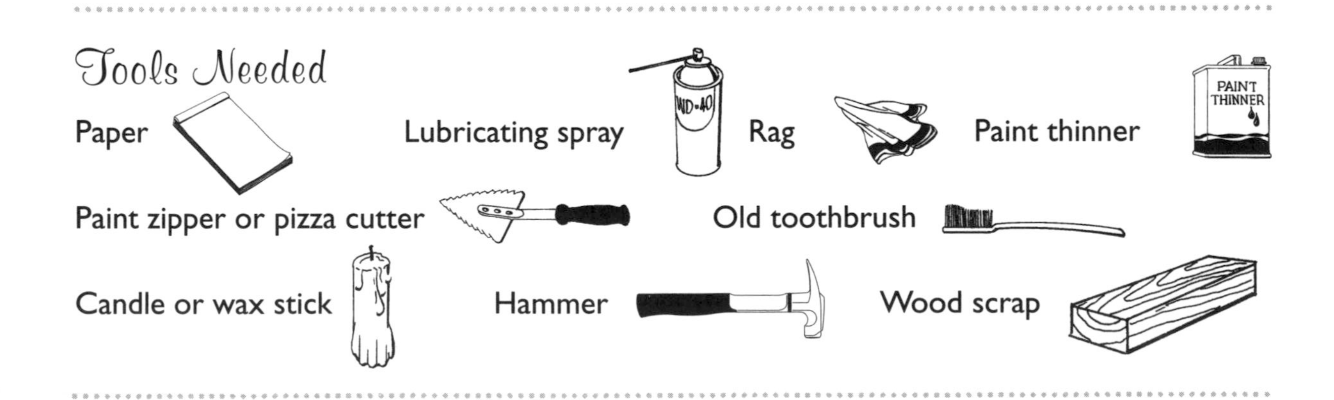

Before starting this task, you'll need to brush up on your window words: sash, sill, stop, channel, and tracks. We all know what the windowsill is, so we'll move right along to sash. The sash is the framework of the window located around the glass. The stop is found on both sides of the window frame. The channel is the area in the window frame where the wood window travels up and down. Tracks are grooves in the sill or sides of the window frame where metal windows travel up and down or side to side.

Finding the Stuck Spot

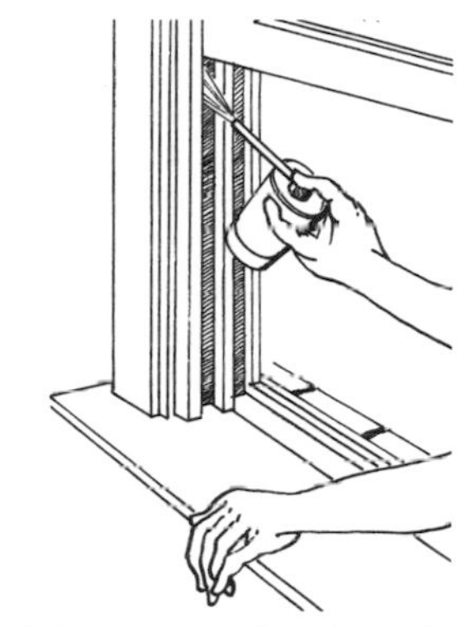
Lubricating window channels

To see where the window is stuck, slide a piece of paper between the sash and the stop, starting from the top of the window and working your way down to the sill. Wherever the paper gets caught is where the window is stuck. Apply lubricating spray to both channels.

Fixing the Problem

Inserting paint zipper in window

Insert the paint zipper or pizza cutter between the sash and the stop, a little bit above the problem spot. Run the tool up and down until it runs smoothly. Perform the same steps on the other side of the window.

If the sides were not stuck, then the problem may lie between the sash and the sill. Insert the paint zipper in this area and run it from one end of the sill to the other.

Try opening the window. If it still won't budge, place the wood scrap over the problem area of the sash and bang it gently with the hammer to jar it a bit.

Lightly hammering on wood scrap

Maintaining Your Windows

Once you're able to open the window, you need to do some preventive maintenance. Pour some paint thinner onto the rag and remove any paint in the channel or tracks that's causing the window to stick. Using a toothbrush, clean the dirt that's accumulated in the tracks (if applicable) and the window channels. Rub the bottom of the candle into the channels or apply lubricating spray to the inside of the tracks to ease the sliding motion. Close the window and try opening it again. If it gets stuck, note the location and repeat the necessary steps.

Replacing a Doorknob

Are your doorknobs falling off? Get a handle on the situation before you get left out in the cold!

Doorknobs (a.k.a. hardware) are classified by function: entry (keyed), privacy (bedroom/bathroom), and passage (hallway/closet). It's important to know these terms so you can walk out of the hardware store with the correct replacement the *first* time.

A doorknob needs to be removed when it no longer closes properly, you want to switch from one type to another (e.g., entry to privacy), or you're replacing a doorknob with a lever (see "Replacing a Round Doorknob with a Lever," page 225).

If an adjustment needs to be made to the latch bolt (i.e., the tubular part of the latch located in the middle of the door hole), check the manufacturer's instructions for realignment.

Tools Needed

Phillips screwdriver 

New doorknob 

Removing a Doorknob

To remove the existing doorknob, use a Phillips screwdriver to loosen the two screws on the cover. It works best to hold down both covers with one hand, while using the other hand to remove the screws.

Removing screws

Installing a New Doorknob

Make sure that the hardware containing the lock will be located on the interior side of the door.

Insert stems of the exterior doorknob into the holes horizontally in the latch case. Place the interior doorknob on the protruding spindle, carefully aligning the stems with the screw holes.

Turn the screws into the holes with one hand, while holding both covers with the other hand. Use the Phillips screwdriver to tighten the screw closest to the door edge first. Tighten the other screw, and test the handle by opening and closing the door.

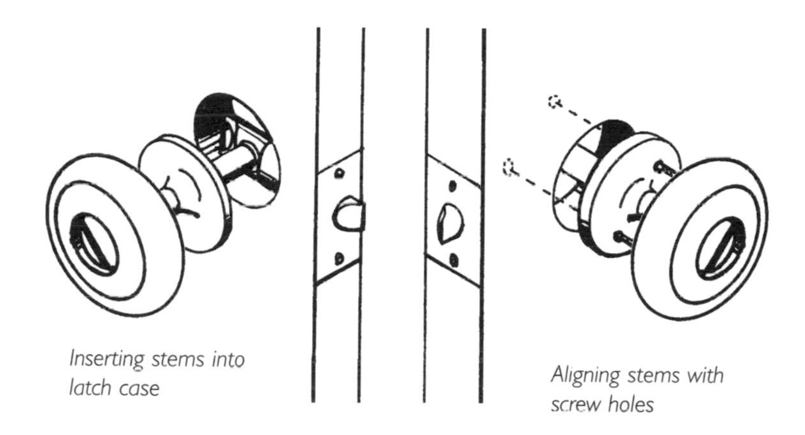

Inserting stems into latch case

Aligning stems with screw holes

Adjusting a Screen/Storm Door Closer

If your screen/storm door is hitting you on the way out, don't take it personally. Simply adjust the closer on the door.

The closer on a screen/storm door regulates the speed at which the door will close shut. You'll need to adjust the closer whenever you make the switch from storm to screen or vice versa because of the weight difference between the glass and screen inserts. It may take a few slight turns of the screw to get the speed to your liking.

The following directions apply to most screen/storm doors with pneumatic closers. If they don't apply to your door, contact the manufacturer.

Adjusting the Speed

To begin, close the screen/storm door. Find the adjusting screw on the pneumatic closer, located at the end farthest from the hinge. Using the screwdriver, turn the screw clockwise to slow the door's speed, and counterclockwise to increase the speed. Open the door and let it close by itself. Continue to adjust the screw until you obtain the desired speed.

Tool Needed

Screwdriver (flathead or Phillips) 

Adjusting screw on pneumatic closer

Maintaining Window and Door Screens

Doesn't it *bug* you when insects are able to find that one itty-bitty hole in the window screen that you thought was too small to be worth fixing! You can put the bug spray away by following this easy repair.

Screens are a great way of letting fresh air in while keeping bugs out, but only if you maintain them by patching holes, no matter the size. Clean the screens before installing, because the first big rainstorm will transfer whatever dirt was on the screens to the windows and doors.

Cleaning Screens

Remove all of the screens and spray with a garden hose. Clean the screens, using a sponge (or wire brush, if applicable) soaked in soapy water. Rinse and dry completely before storing or placing in windows and doors.

Cleaning screens

Tools Needed

Sponge 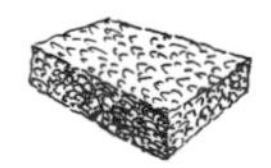 Garden hose or bathtub/utility sink 

Wire brush (for metal screens)  Bucket with warm soapy water 

Patching Screens

If you're particular about the appearance of uniformity, this repair will not work for you because a patch on your screen will be noticeable and you'd probably be happier replacing the entire screen. If uniformity isn't important, you can purchase screen patches in hardware stores.

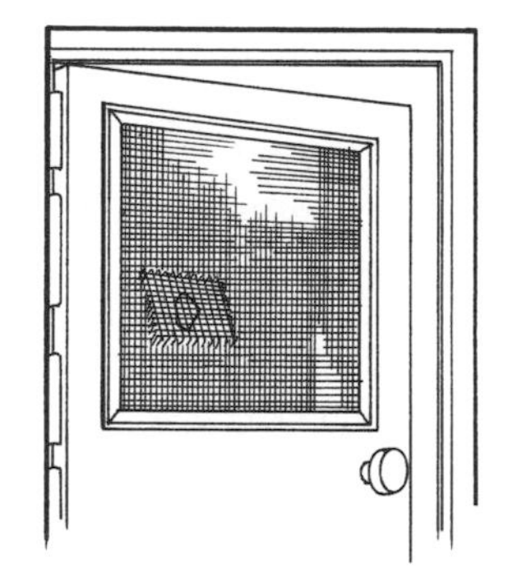

Patch installed in door screen

Place the patch on the interior side over the small hole and feed the wires into the screen until all four sides of the patch are installed. If you need more wire to get a better hold on the screen, remove one wire from each side and then feed in all sides. If you want the patch to be smaller, use the scissors to cut it to the desired size (be sure to cut the patch over a garbage can because it will make a mess).

Spline is the name given to the rubber rope that keeps the screen in place inside the window or door frame.

Replacing Screens

You'll need to replace a screen if it has a tear rather than a hole. But if a screen sags, you may be able to reuse it. If you want to do that, follow the instructions below, using new spline with the original screen. Otherwise, purchase new screen.

Tools Needed

Screen patch repair kit (contains small mesh squares)

Scissors

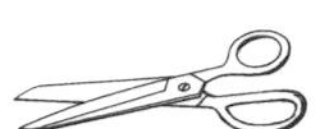

Before you start, there are some things you need to know about this repair: 1) you can do it yourself, but having an extra pair of hands is helpful; 2) the diameter of the new spline has to be the same as that of the old spline; and 3) it works best to lay the screen frame on the floor, with one end pushed up against a wall.

Removing an Old Screen

Remove the screen frame (wood or metal) from the window or door by snapping it out or using a screwdriver to loosen the screws. Pry the spline from the perimeter of the screen with the screwdriver. Take the old spline to the hardware store so you can purchase exactly the same size.

Removing spline

Tools Needed

Screwdriver (flathead or Phillips)  New screen cloth 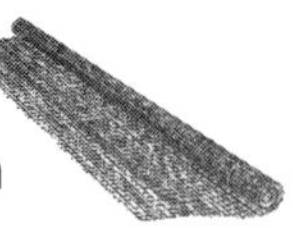

Scissors  Spline tool (concave and convex wheel) 

Spline  Utility knife 

Pull out the old screen and discard. Put the new screen cloth over the window or doorframe, leaving a minimum 1/4-inch overlap. Cut all four corners at an angle so the screen won't bulge.

Securing a New Screen

Beginning at one corner, push the convex wheel down on the screen and into the channel (i.e., groove) making an indent in the screen as you go up and down one side. Take the spline and insert it on top of the screen that's in the channel. Now use the opposite end of the spline tool (the concave wheel) to push the spline into the channel. To ensure that the screen won't be too tight, rest one hand on the center of the screen. If the concave wheel doesn't work, you may need to use the screwdriver. Proceed with the other three sides.

Once all of the sides have been done, push the spline into the four corners with the screwdriver. Use scissors to cut away any large pieces of screen and then a utility knife to carefully trim the excess close to the frame.

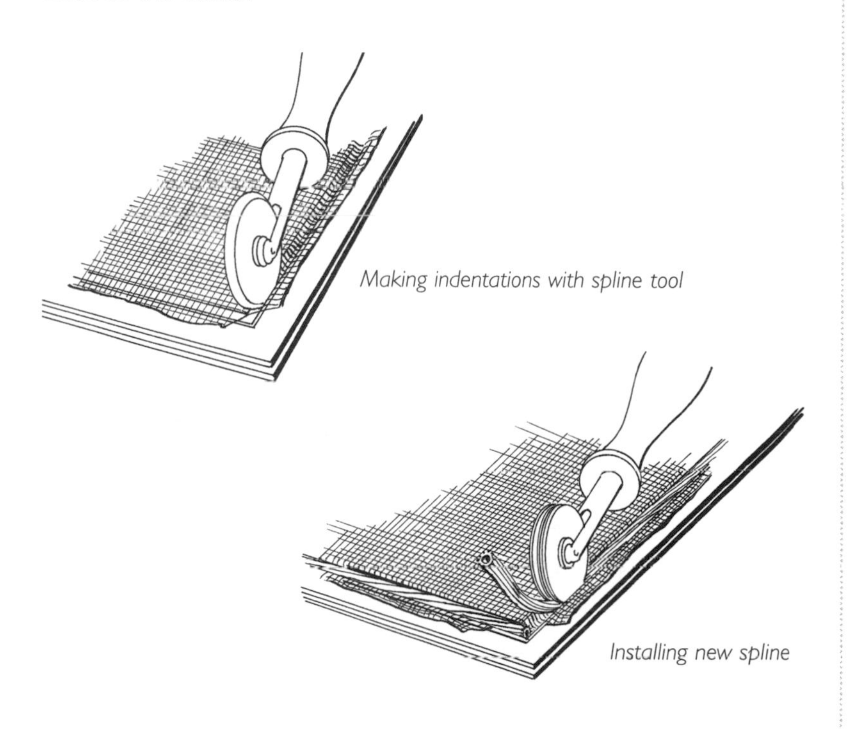

Making indentations with spline tool

Installing new spline

Home Safety

Fire Security

- Smoke Detectors
- Carbon Monoxide Detectors
- Fire Extinguishers
- Planning an Emergency Escape Route
- Safety Guidelines for a Fireplace/Woodstove/Portable Heater

Home Security

- Installing Window Pins in a Double-Hung Window
- Installing a Door Peephole
- Securing a Sliding Glass Door
- Resetting an Automatic Garage Door Opener

Safety Measures

- Performing Garage Door Safety Tests
- Adding Safety Devices to Window Treatments
- Installing Interior Window Guards
- Freeing Someone Locked Inside a Room
- Setting a Water Heater's Temperature
- Installing a Handheld Shower Unit
- Replacing a Round Doorknob with a Lever
- Preventing Falls
- Practicing Ladder Safety
- Patching Cracks in a Driveway
- Safety Checklist for Children and Seniors

We've all been blessed with 20–20 hindsight *after* an accident happens. Even though no one can predict the future, there are things you can do today to ensure a safer tomorrow.

Don't put blinders on when it comes to your safety and that of those in your care. By following the repairs and safety tips we provide in this section, you can turn your home into a safe haven.

When it comes to home safety, the number one rule should be: *don't assume your home is safe*. Don't assume that your smoke detector is working, *know it*. Don't believe that your sliding glass door is adequately secured, *know it*. Don't guess that your water temperature is 120°F, *know it*. Install the safety devices suggested in this section and properly test them, as recommended.

Fire Security

Smoke Detectors

Never take batteries out of another product—use only new batteries.

Sometimes it takes a small child to teach you a big lesson. Amy's six-year-old daughter, Caroline, came home from school one day and said that a firefighter had come to her class. The firefighter asked the children, "What's the first thing you do when you hear the smoke detector go off?" to which Caroline replied, "Hit it with a pot holder!" The firefighter was surprised by the answer and asked her to explain. She said, "Every time my mommy burns something in the kitchen, the smoke detector goes off and she hits it with a pot holder." Now Amy has a smoke detector specifically designed for the kitchen. Shame is a powerful tool!

There are three types of smoke detectors: 1) battery-operated; 2) hardwired; and 3) hardwired with battery backup. Unless you're the original homeowner, you may not know whether your home has more than one type. Therefore, it's important to check every smoke detector in your home.

Identifying Smoke Detectors

Battery-operated smoke detectors receive their power from batteries. Unfortunately, more than a third of all battery-operated smoke detectors are not working because the batteries are either missing or dead.

A hardwired smoke detector receives its power through the home's main electrical supply. The disadvantage of having a hardwired smoke detector is that if there is an electrical outage in your home, the smoke detector will not work because its only source of energy has been shut off. The solution is to purchase a hardwired smoke detector with a battery backup system.

So, if the battery-operated and the hardwired with battery backup smoke detectors both use batteries, how can you tell which one you have? You can't look for wires because the hardwired smoke detector's wires are concealed behind drywall. The following test is a simple way to find the answer.

Before testing any of the three types of smoke detectors, place the stepladder underneath the smoke detector and take the top off.

Most fire departments across the country have programs that provide smoke detectors for free or for a nominal charge to low-income families. Contact your local fire department for more information.

Identifying a Battery-Operated Smoke Detector

Remove the battery and place the top back on the smoke detector. Press the test/silence button (found on the outside of the detector). If the alarm does not go off, then it is a battery-operated smoke detector, because you removed the smoke detector's only source of power.

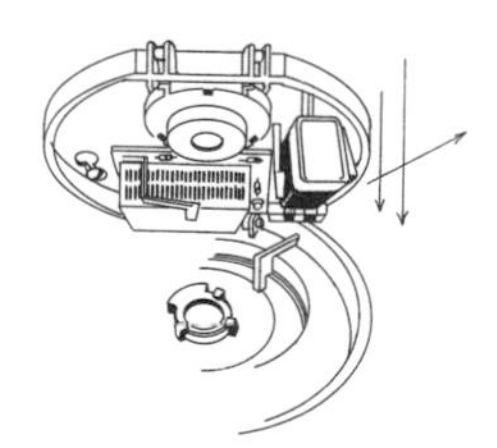

Battery-operated smoke detector

Disconnected battery

Depressing test/silence button

Tools Needed

Stepladder or ladder

New 9-volt battery

If you don't see a battery, don't assume that the smoke detector is hardwired—the battery could be missing. Look for a battery connector, and if you find one, place a new battery in it immediately.

Identifying a Hardwired with Battery Backup Smoke Detector

A hardwired with battery backup smoke detector will have a battery connector. If there is a battery, remove it. Replace the top on the smoke detector. Press the test/silence button (found on the outside of the detector). If the alarm sounds, then the smoke detector is hardwired with a battery backup, receiving its power from two electrical sources. If the smoke detector does not have a battery, place a new battery in it immediately.

If your home has hardwired or hardwired with battery backup smoke detectors, all of the alarms should go off at the same time.

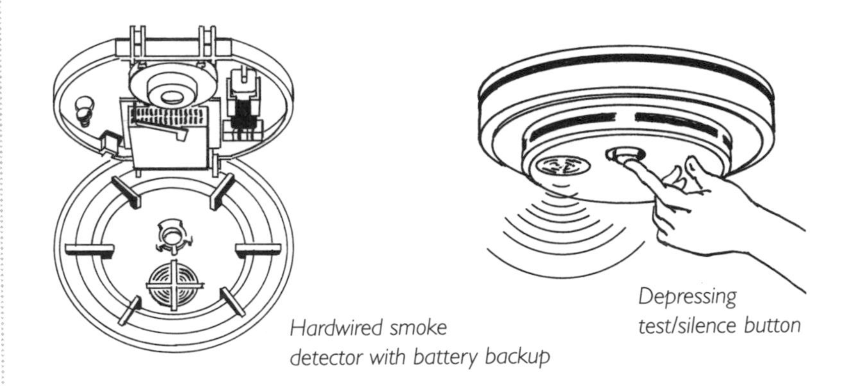

Hardwired smoke detector with battery backup

Depressing test/silence button

Identifying a Hardwired Smoke Detector

A hardwired smoke detector will *not* have a battery connector. Place the top back on the smoke detector and press the test/silence button (found on the outside of the detector). If the alarm goes off, it signals that the smoke detector is hardwired because its only source of power is the home's electrical system.

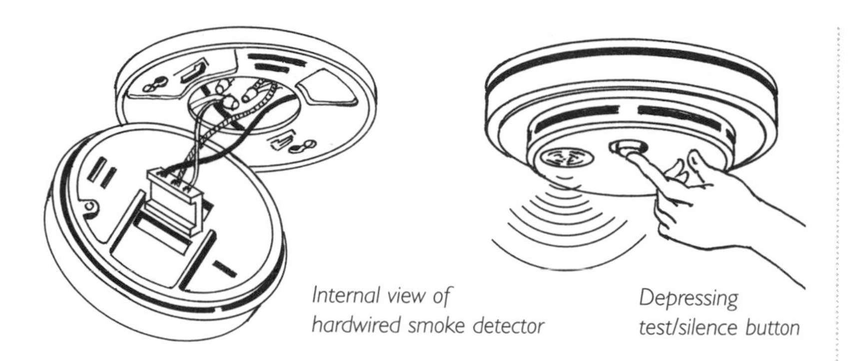

Internal view of hardwired smoke detector

Depressing test/silence button

Where to Install a Smoke Detector

The International Association of Fire Chiefs recommends you check that your home has an adequate number of smoke detectors. There's no magic number, just a rule: there should be smoke detectors on every level of your home (including the basement) and one in front of each bedroom.

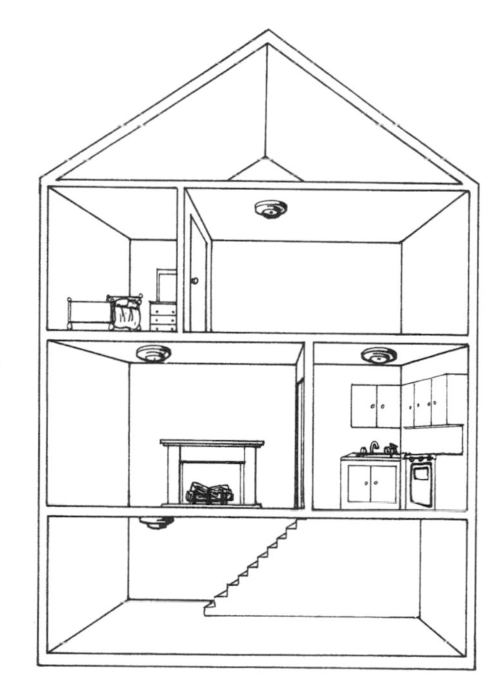

Locations for smoke detectors

Smoke detectors come in a wide range of choices from several manufacturers, often with extra features, such as a light, which may require additional batteries.

Basement

- Install a smoke detector at the bottom of the landing, not at the top.
- Never place a smoke detector near the furnace.

Kitchen

- Install a smoke detector specifically designed for kitchen use.

Living Room

- Smoke detectors should be mounted on the ceiling in the middle of the room and at least 4 inches from the corners.
- If a smoke detector has to be mounted on the wall, install it 4 to 12 inches from the ceiling and as close to the center of the room as possible. Be sure there are no obstacles that could obscure the path of rising smoke (e.g., a picture underneath the smoke detector).
- Keep smoke detectors away from fireplaces and wood-stoves; these could cause false alarms.

Bathrooms

- Install smoke detectors 10 feet from bathrooms because excessive steam can set them off.

Bedrooms

- There should be a smoke detector outside of each bedroom.
- If the bedroom hallway is longer than 30 feet, there should be a smoke detector at each end.
- If you sleep with the doors closed, place a smoke detector inside the room as well as outside.
- Someone who has a hearing disability should purchase a smoke detector that uses lights or vibrating devices.

Installing a Battery-Operated Smoke Detector

Location, location, location. It's the most important thing to consider when installing a smoke detector (P.S. Feng shui cannot enter into the equation). Before installing a smoke detector, you need to read the manufacturer's instructions. Most list a drill as one of the tools needed to install a detector, but don't worry—we offer an alternative for those of you who don't own one.

Remove the top of the new smoke detector, either by lifting or twisting it off. Inside you should find a new battery (batteries) and screws.

The Consumer Product Safety Commission states that in 20 percent of households with at least one smoke detector, the alarm doesn't work because of a dead or missing battery.

Marking the Location

Move the stepladder to the spot where you'll be installing the detector. Climb up wearing the tool belt containing the pencil, base, and top of the detector, hammer, screwdriver, mounting screws, and drywall anchors. Place the base of the smoke detector against the ceiling (or wall) and use the pencil to mark the centers of the two mounting holes. Temporarily place the base on top of the ladder.

Marking location of new smoke detector on ceiling

Tools Needed

New smoke detector (with *new* battery) 

Stepladder or ladder 

Tool belt 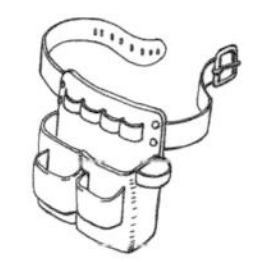

Pencil (sharpened) 

Hammer 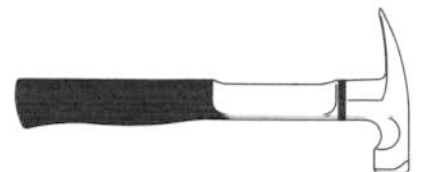

Screwdriver (flathead or Phillips) 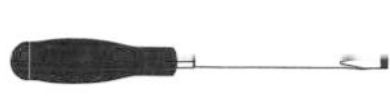

Self-drilling drywall anchors 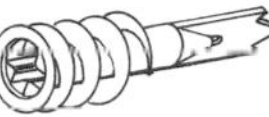

Mounting a Smoke Detector

Hammer the tip of each of the drywall anchors into the designated spots. Using the screwdriver, screw the anchors until they're flush with the wall/ceiling. Then screw the mounting screws into the anchors. Line up the base, rotating it around until it fits securely. Connect the battery (batteries).

Inserting mounting screws into anchors

Replace the cover. A newly installed smoke detector may beep when it's first connected, but you still need to test it. Press the test/silence button (found on the outside of the detector) and hold for 5 to 10 seconds to activate alarm. If you installed the smoke detector correctly, the alarm will sound. If the alarm fails to go off, check the battery's connection or its expiration date.

Testing a Smoke Detector

It's important to remove dirt and dust because they can impede the detector's sensitivity to smoke. Gently vacuum the cover every month.

Move the stepladder underneath the smoke detector and remove the cover. If the smoke detector is battery-operated or hardwired (i.e., connected to the electrical wiring of your home) with a battery backup, check for a battery inside. If your smoke detector has a light, then it should have one or more batteries for the light, along with the 9-volt battery for the alarm. Replace the cover.

Tools Needed

Stepladder or ladder

New 9-volt battery

Hard-and-Fast Rules

Does your smoke detector get tested only when you burn something in the kitchen? You should test your smoke detectors every month and after a vacation. Change the batteries when the smoke detectors emit a beeping sound; even if they don't beep, change them every fall, or use your birthday as the day to remember to make the change. Never replace a dead battery with a used one, even if it still works. If your smoke detector is 10 years old or more, replace it with a new one that is an Underwriters Laboratories (UL) listed model.

Press the test/silence button (found on the outside of the detector) for 5 to 10 seconds to activate alarm, then let go. If the alarm doesn't go off, replace the battery. If the smoke detector is hardwired, all the hardwired smoke detectors in your home should go off at the same time because they're all on the same circuit branch. If they don't, check the circuit breaker and/or fuse before calling a certified electrician.

Taking the top off a smoke detector

Carbon Monoxide Detectors

Do you have headaches? Are you tired, nauseated, and confused? Do you get dizzy spells and become irritable? We could go for the cheap laugh and say, "Yeah, once a month," but this is no laughing matter. These are the signs of carbon monoxide (CO_2) poisoning, and they need to be taken seriously. The Consumer Product Safety Commission states that every year approximately 7,000 people are admitted to emergency rooms and 200 people die from this odorless, colorless poison. If you or anyone in your home is experiencing these symptoms, leave immediately and call 911.

There are two things you need to do to keep this from happening to you: 1) have your chimney, furnace, and all oil and gas appliances in your home inspected each year; and 2) purchase carbon monoxide detectors.

Purchasing a Carbon Monoxide Detector

There's a wide selection of detectors on the market, ranging from plug-ins to combination smoke and carbon monoxide detectors, and from hardwired detectors to those designed to sit on a shelf. No matter which you choose, make sure that it has the Underwriters Laboratories (UL) seal. If you buy a detector that's hardwired, get one with a battery backup so it will still work in case of a power outage.

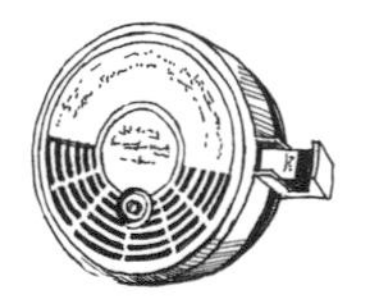
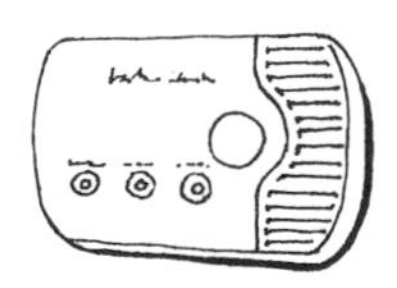

Carbon monoxide detectors

Maintaining a Carbon Monoxide Detector

Test and clean the carbon monoxide detector monthly, and if it's battery-operated, replace the battery every year.

If the carbon monoxide detector is mounted on a wall, move the stepladder underneath it.

Gently vacuum the exterior of the detector and its surroundings. Press the test/silence button (found on the outside of the detector) to activate the alarm, for 5 to 10 seconds. The alarm should go off. If the detector is battery operated and the alarm doesn't sound, replace the battery. If it's hardwired and the alarm doesn't go off, check the circuit breaker/fuse before calling a certified electrician.

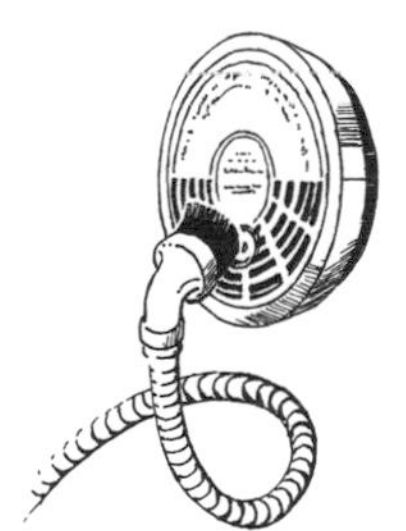

Vacuuming carbon monoxide detector

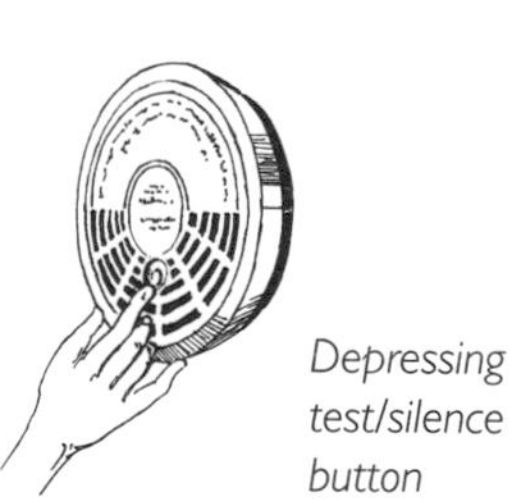

Depressing test/silence button

It's important to have at least one carbon monoxide detector in your home, preferably one on each level. It should be located at least 15 feet away from the gas or oil furnace or water heater, and in the hallway outside of a bedroom.

Tools Needed

Stepladder or ladder

Vacuum with brush attachment

New 9-volt battery

Fire Extinguishers

Leigh had a small fire in her kitchen and thought, "No problem. I'll just get my fire extinguisher, and *presto,* I'll put the fire out." But as she picked up the extinguisher, she realized she didn't know how to use it. In fact, she needed to read the instructions first! Unfortunately, the fire didn't wait for Leigh to get her act together.

If a small fire broke out in your kitchen, would you have to read the instructions on the fire extinguisher to know how to use it? Would you need to find your reading glasses first?

Most small fires begin in the kitchen, but it takes only a matter of minutes for a fire that's controllable to get out of control. There are some things Leigh could have done prior to and during the accident that would have changed the outcome greatly: 1) She could have prepared herself by *test-driving* a fire extinguisher prior to a fire. In a panicky situation, it's hard to concentrate and follow directions; 2) Leigh didn't call 911 because she assumed incorrectly that she had the fire under control; and 3) she forgot to turn off the stove, where the fire originated!

Purchasing a Fire Extinguisher

Before you learn how to operate a fire extinguisher, it's important to know which type you need for your home. Fire extinguishers come with ratings for the size and type of fire they can put out. For example,

Type A is for ordinary combustibles such as paper, wood, cloth, and some plastics. Type B is for flammable liquids, such as grease, oil, flammable gas, and oil-based paint. Type C is for electrical equipment such as appliances, televisions, circuit breakers, and fuse boxes. The number in front of the letter indicates the size of the fire it can extinguish; a larger number indicates it can cover a bigger area.

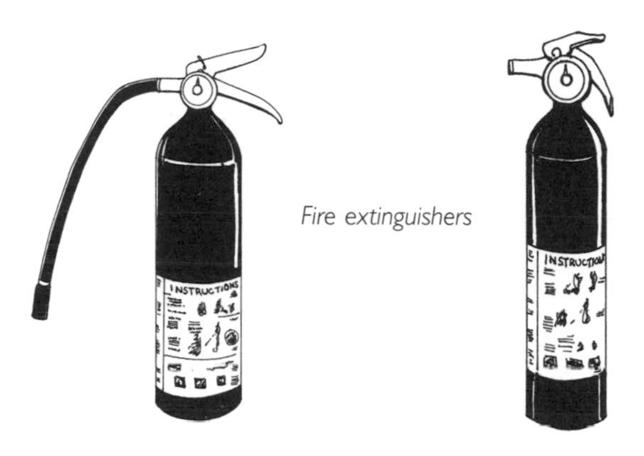

Fire extinguishers

The most practical household fire extinguisher to own is one with the following rating: 1A; 10B; C or 3A; 10B; C. These ratings will allow you to put out most small fires. There are fire extinguishers specifically designed for garages, recreational vehicles (e.g., boats), and kitchens. Because 65 percent of home fires start in the kitchen, it's wise to have a fire extinguisher in that room.

Using Your Extinguisher

There's a simple acronym to remember when using a fire extinguisher: **P.A.S.S.**

P ull the pin
A im at the base of the fire
S queeze the handle or trigger
S weep in a side-to-side motion

If you've never used a fire extinguisher before, we suggest you buy one for practice. The compound is released with such great force that you can easily lose control in the first few seconds. However, we're not suggesting you start a fire so you can use the fire extinguisher! Just go outdoors, or use a sink for target practice.

After you've put out the fire, the compound needs to be cleaned up immediately before you use any electrical appliances in the area or turn the electric power on, because the compound can conduct electricity when wet. Also, the extinguisher needs to be completely emptied before you throw it out. Contact your local fire department for information on how to properly dispose of the extinguisher.

Maintaining Your Fire Extinguishers

Check that a fire extinguisher is full by noting the location of the arrow in the gauge, found on the front of the fire extinguisher. The green section denotes "Full" (or "Overcharged") and the red section states "Empty" (or "Recharge"). Use the fire extinguisher only if the arrow is in the green section, and if it isn't, then throw the extinguisher away.

Clean away any cobwebs or spiders that may be inside the nozzle that could clog it. Check that the pin is properly secured. If the pin has been removed and replaced, assume that the extinguisher has been discharged and throw it out. Perform this monthly maintenance on every fire extinguisher in your home.

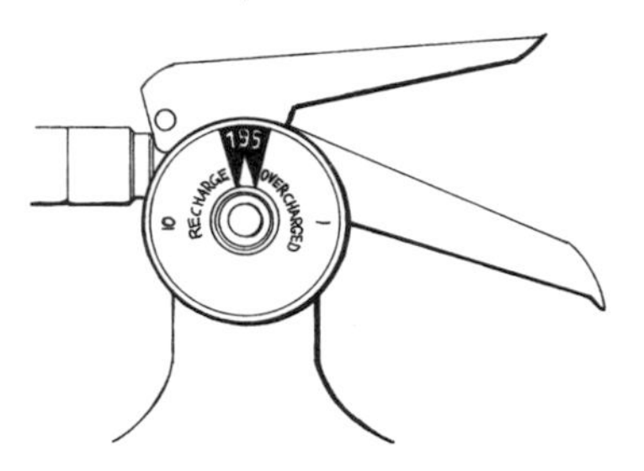

Indicator on fire extinguisher

Tools Needed

Rag 

Fire extinguisher 

Planning an Emergency Escape Route

When Ann proposed to her family the idea of practicing an emergency escape route, she received a silent standing ovation; in other words, everyone got up to leave. Knowing how competitive her family was, she decided to turn her idea into a contest of who could get out of the house first. After practicing the escape plan a few times, the family was able to determine who was the fastest and who needed assistance. This was one family event that made everyone a winner.

Developing an emergency escape plan requires the full cooperation of everyone living in your home. Practice it twice a year. No one gets a homework pass—even if you live alone!

You'd be amazed at how your brain doesn't work during a fire. Therefore, the emergency escape plan needs to become rote to you and everyone in your home. Don't take for granted that someone will think to get all of the young children out, or remember to help an elderly parent to safety. How about your baby-sitter—does she know the escape routes and checkpoint? These are things you need to consider when designing the escape plan.

Tools Needed

Pencil and paper 

Smoke detector 

Watch 

To prepare yourself and your family for a fire, rehearse crawling out of a room, test a door and doorknob for heat before opening, learn to *stop, drop*, and *roll* if your clothes catch on fire, and practice releasing safety locks on windows and doors.

Identifying Escape Routes

First, have everyone go to separate rooms with pen and paper and draw at least two escape routes (e.g., windows and doors). Check for items blocking a path to the exits, locks that are too high for children, doorknobs that are not user-friendly for seniors, and windows that are painted shut or have security bars. Are the fire escape ladders readily available for use? Make notes about problem areas and remedy them quickly.

Next, take everyone outside to pick a location where they'll congregate once they've exited the house.

Identifying escape routes

Performing a Drill

Now you're ready to test yourself and/or your family. Perform the first safety drill at night with the lights out. Have everyone lie in their bed to wait for the alarm to go off. Look at your watch and note the time. Press the test/silence button (found on the exterior of the detector) on the smoke detector outside of your room, go back to your bed, then find the exit. Once everyone is at the checkpoint outside, look at your watch. How long did it take for everyone to get out? What problems did you or others have in reaching the exits? Were the safety ladders easy to use? Was it difficult for children or seniors to turn the doorknobs or open windows? Did the meeting spot work, or would another location be better?

Depressing test/silence button

Try the emergency escape plan again, but this time tell everyone to pretend the rooms are filled with smoke and the visibility is low. Tell them to crawl on the floor where the air is cleaner and cooler. How did everyone do? How long did it take? Write notes and make any necessary adjustments. Remember, practice makes perfect.

Safety Guidelines for a Fireplace/Woodstove/Portable Heater

For all you queens of clean, here's the real dirt on how to remove ashes from a fireplace: you can't, at least not right away. If you don't believe us, ask Carol Ann, a certified perfectionist who vacuumed the ashes too early from the fireplace, only to have them ignite inside her vacuum cleaner.

The rising cost of fuel has made some people find additional or alternative ways to heat their homes, such as fireplaces, woodstoves, and space heaters. Before you purchase any heating equipment, make sure it is safety-listed by the Underwriters Laboratories, and ask your local fire department for your county's fire and building codes. For example, some counties ban the use of portable kerosene space heaters because they pose a serious fire hazard. And we can't say it enough—it's extremely important to have working smoke and carbon monoxide detectors throughout your home.

Fireplaces

Ashes can contain embers for up to a week after a fire. But if you use your fireplace or woodstove consistently during the winter months, you can't wait a week before removing the ashes. Therefore, the safest way to perform this task, no matter how long after a fire, is to use a metal scoop for cleaning the fireplace and a metal garbage can (with a lid) to store the ashes. Never store the metal garbage can in the house, on a wood deck, in a garage, or near flammable materials, and don't throw your ashes into the woods. Instead, wait until spring, then recycle them into your soil.

Fireplaces should be serviced by a professional chimney sweep every year prior to winter use. The chimney sweep will:

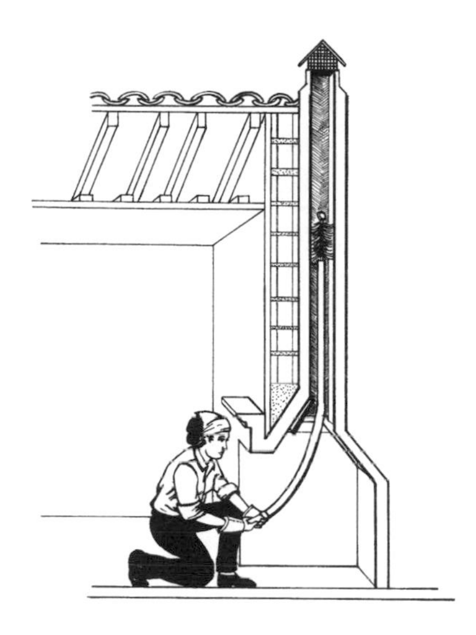

Fireplace serviced by professional chimney sweep

- **Check for internal and external signs of damage.**
- **Sweep and vacuum the smoke chamber and shelf.**
- **Remove creosote, a chemical substance that forms when wood is burned and is deposited in the lining of the chimney. A chimney fire can start with as little as ¼-inch buildup of creosote.**

You can also help prevent fires by using a fireplace screen, burning only seasoned hardwood, checking that the damper is open before starting a fire, and never using the fireplace as a cooking appliance or incinerator.

Woodstoves

Owners of woodstoves should follow the manufacturer's instructions for proper care and usage. Basic rules are:

- Burn only seasoned hardwood.
- Never use the woodstove as a cooking appliance or incinerator.
- Place the woodstove on an approved floor protector or fire-resistant floor.
- Have it serviced by a professional chimney sweep yearly.

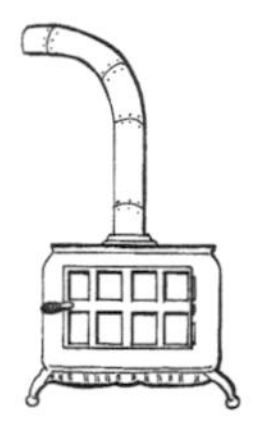

Space Heaters

Space heaters, electric or fuel-operated, are a popular heat source because they're inexpensive and don't take up much space. No matter which kind of space heater you have, the list of *do's* and *don'ts* is long:

- Keep space heaters at least 3 feet from clothing, furniture, wallpaper, pets, and people.
- Never leave a space heater unattended.
- Turn off a space heater before going to sleep.
- Do not use the space heater for drying clothing.
- Read the manufacturer's operating manual.
- Do not use the space heater for cooking.
- Place the space heater on a level surface.
- For fuel-powered space heaters, only use fuel specified by the manufacturer, store the fuel outside in an approved container labeled correctly, and refuel the space heater outdoors after it has cooled.
- For electric space heaters, keep the cord knot-free, don't cover them with anything, and keep them away from water. Check for damaged cords and plugs, and don't use extension cords.

Home Security

Installing Window Pins in a Double-Hung Window

Maureen thought she was the 90s version of Mary Tyler Moore. Having just graduated from college, she got a job at a news station and an apartment in an old Victorian house. As a single woman in a new city, Maureen knew she had to be realistic about her surroundings, so she talked her landlord into allowing her to install window pins. The next day at work, she suggested to her boss that they do a feature on how women can make their homes safe. After receiving a big pat on the back, Maureen smiled and thought, *I just might make it after all!*

Sorry ladies, but there's no way to avoid using a drill for this project because you need to bore a hole into two windows. Note: A cordless drill is heavier because it contains a battery pack, but you won't have to mess with a cord.

A double-hung window has a crescent latch designed to keep the upper and lower windows together, but the latch is *not* a safety device. The best way to protect against an intruder is to secure double-hung windows with window pins. An inexpensive and effective alternative to locks, window pins are designed to keep the lower window from being raised. Don't be alarmed—you can still get fresh air at night by

Double-hung window

drilling a second set of holes in the sash. Just be careful not to drill too low or your window will be raised too high, creating easy (and unwanted) access.

Avoid installing window pins that require a key or tool for removal so that in the event of a fire, you can quickly pull the pin from the window and exit.

If you can't find traditional pins with permanently installed release devices, use 12 d common nails or 5/16-inch diameter eyebolts.

Preparing the Window

Close the window and secure the latch. Tear off two pieces of masking tape and stick them to the upper right and left corners of the inside sash. Use the tape measure to find the centers of the corners and mark with a pencil. (The masking tape will also help protect a wooden sash from splintering when you drill a hole.)

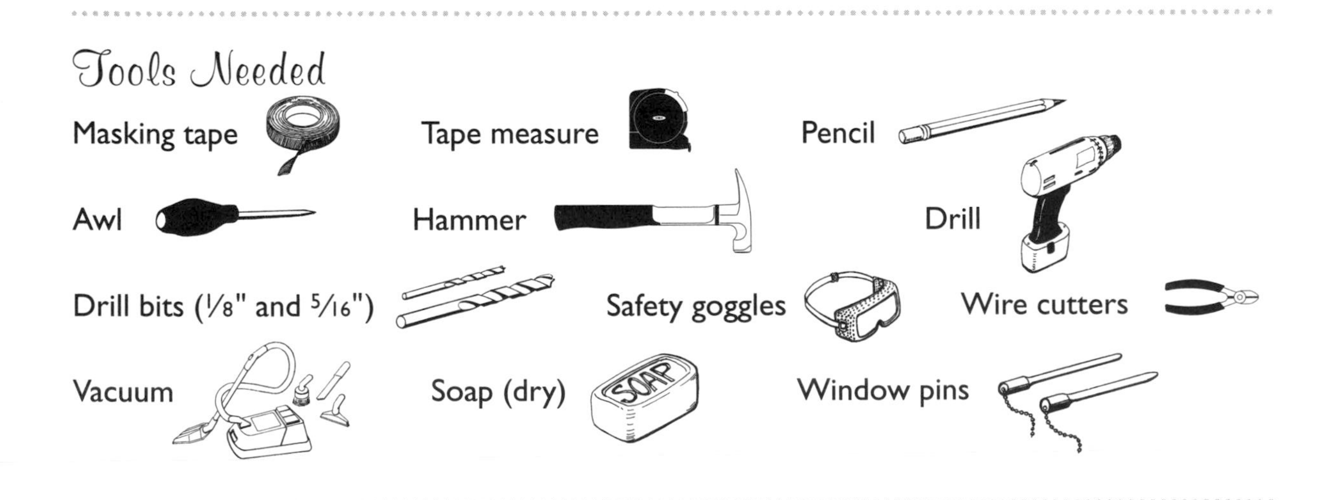

Drilling the Holes

Use the awl and hammer to create an indentation on the marked spots on the taped corners. Insert the drill bit into the drill just enough so that the chuck (the round part of the drill that houses the bit) securely grasps the drill bit.

An awl and hammer are used to create a starter hole that will make it easier to insert a drill bit, screw, or nail.

Creating indentations with awl and hammer

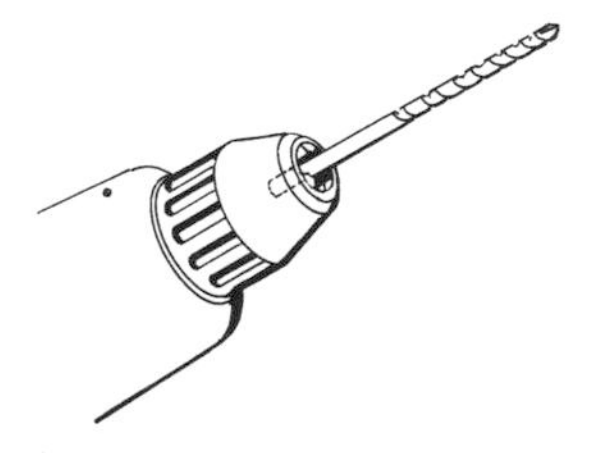

Inserting drill bit

It's important not to drill completely through both sashes because the pin will then be accessible to a burglar from the outside. An easy way to keep from drilling too far is to lay the drill bit on top of the sashes (with the pointed tip facing the window) and place a small piece of masking tape on the drill bit to mark where you need to stop.

Placing drill bit on top of sash

Put on the safety goggles and drill a starter hole using the smaller drill bit. Firmly grasp the drill with both hands and apply moderate and even pressure while drilling at a slight angle (about 20 degrees) through the inside sash and three-quarters of the way through the outside sash.

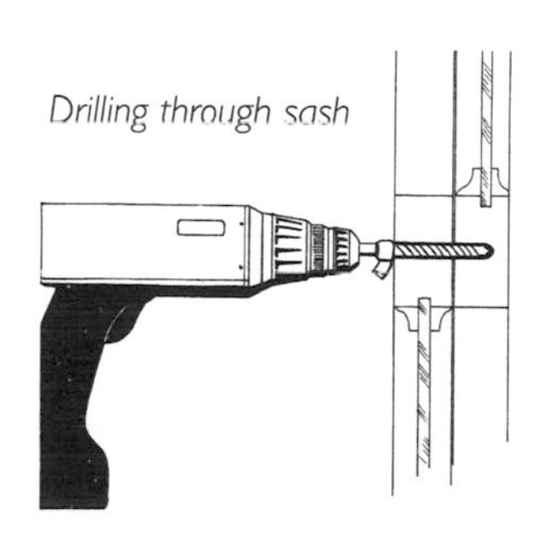

Drilling through sash

After drilling your initial hole, use the larger drill bit to drill through the original hole. Don't be alarmed if you experience some resistance, because you may be drilling through the weather seal (usually made of aluminum, copper, or plastic). Clean the hole by drilling in and out a few times.

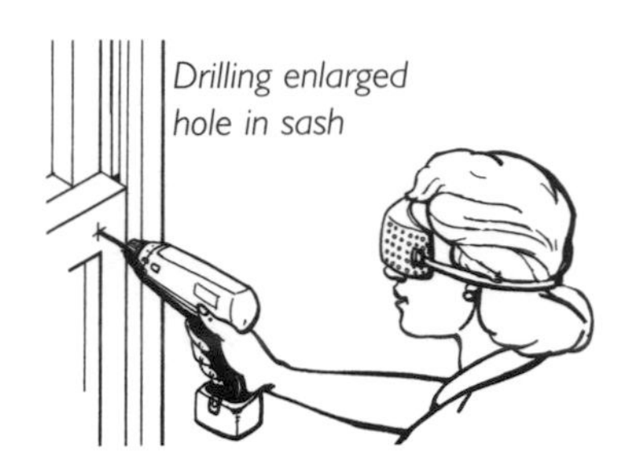
Drilling enlarged hole in sash

Installing Window Pins

Use wire cutters to remove any rough edges from the outer sash. Take off the safety goggles.

Remove the masking tape from the corners of the window. Vacuum the holes and the surrounding area to eliminate excess wood or metal fragments. Use the soap to coat the pins before inserting into the drilled holes, and check to be sure that the pins are a little loose for easy removal.

Repeat this procedure on all double-hung windows in your home, including the upper-level windows, which most people leave unsecured, thereby providing easy access for burglars.

Installing a Door Peephole

"A*ren't you afraid to live alone? Why don't you get a dog?*" Satsu's mother always asked. Satsu thought the best way to somewhat appease her mother would be to add security devices to her home. So, she contacted her local police department and had a policewoman come over to explain how to improve security. The next time her mother visited, Satsu looked out the peephole, but before undoing the deadbolt lock, asked her for two forms of I.D.

A peephole is a device for an exterior door that provides security by allowing the homeowner to see out without being seen. Peepholes should be installed in every solid exterior door (including the door connecting to the garage) where sight is restricted.

When purchasing a peephole, check the package for details such as the size of the drill bits needed, the diameter, the size of door (i.e., width) it will fit, and the field of view. The field of view is very impor-

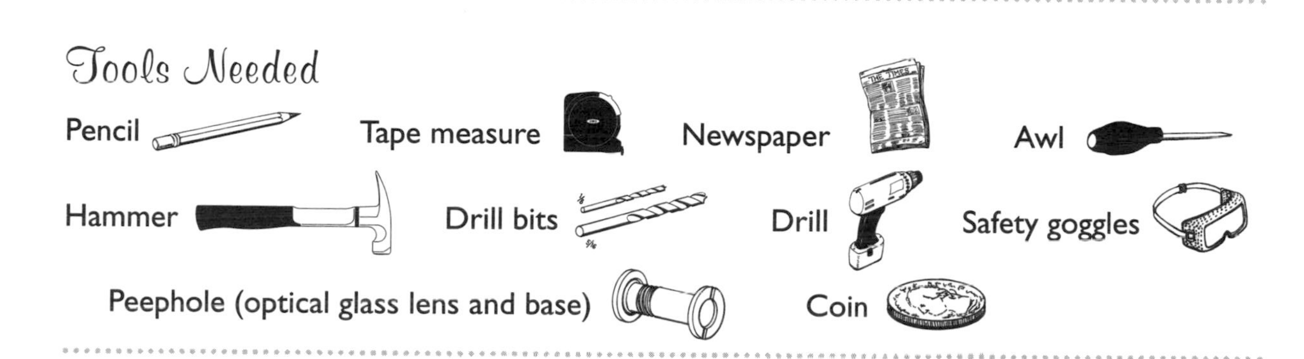

tant because it determines how much you can see. Your peephole should have a field of view of no less than 190 degrees.

Preparing the Door

Mark with a pencil where you want the peephole to be placed, making sure it's at a comfortable height for you. Use a tape measure to find the center of the interior side of the door and adjust your marking if necessary.

Put newspaper underneath the door (you're going to be making a mess). Use the awl and hammer to create an indentation in the door to make it easier to insert a drill bit.

Drilling the Hole

Note: It is extremely important to keep the drill straight while drilling.

Insert the smaller drill bit into the drill just enough so that the chuck (the round part of the drill which houses the bit) securely grasps the drill bit. Wearing the safety goggles, insert the drill bit into the indentation and drill a starter hole completely through the door. Using the larger drill bit, drill a hole halfway through the interior and exterior sides of the door. Remove the goggles.

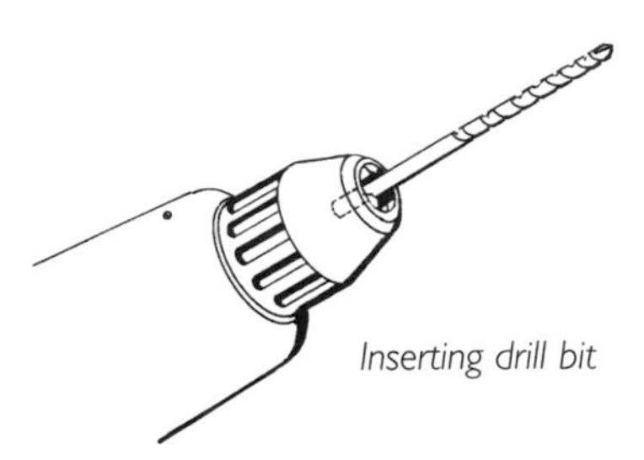

Inserting drill bit

Drilling hole into door

Installing a Peephole

Coin inserted into peephole

Disassemble the peephole by unscrewing the optical glass lens from its base. Install the lens from the exterior side of the door and screw the base onto it from the interior side of the door with your fingers. Tighten it by placing a coin into the slots on the interior side of the peephole.

Securing a Sliding Glass Door

Did you know that 80 percent of burglars enter through a door? Take the "Welcome" mat away by securing your sliding glass doors correctly.

Don't be fooled into thinking that the lock that came with the sliding glass door is enough to protect you. A burglar can gain entry by breaking the glass, prying the door to slide it open, or lifting the door out of its track.

In addition to the door's original lock, the best security system for your sliding glass door is a triple threat; 1) removable drop bar; 2) track screws; and 3) foot lock. You can also purchase safety devices specifically made for sliding glass doors—the options are yours.

Removable Drop Bar (Charlie Bar)

A removable drop bar, also known as a Charlie bar, is more of a visual than a physical deterrent. A burglar can pop the bar open, but chances are he'll opt to go to a home that doesn't have one. Never use a broomstick or other piece of wood in lieu of a removable drop bar because it can be broken under pressure.

Preparing the Door

Close the sliding glass door. You'll want to install the bar at about half the height of the door, so measure the door from top to bottom and mark the middle point with the pencil.

Temporarily place the rotating bracket in the doorjamb of the stationary door and the locking saddle on the inside edge of the sliding door and note where the screws will be inserted. Make sure the bar is straight by using a level. (For something to be level, the bubble has to be centered between the two black lines.)

Drilling the Holes

Use the awl and hammer to make indentations on the pencil marks. (You're making a starter hole so that it will be easier to insert the drill bit.) Insert the drill bit into the drill just enough so that the chuck (the round part of the drill that houses the bit) securely grasps it. Put on the safety goggles and drill a hole into each of the indentations.

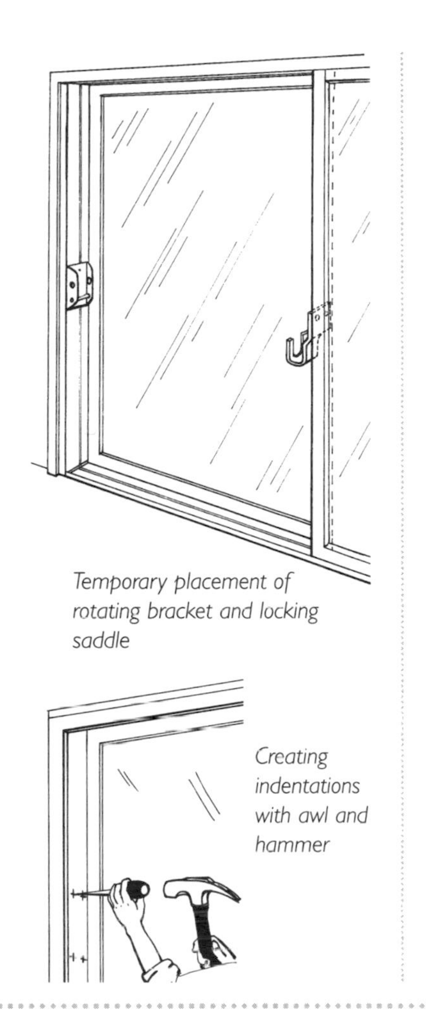

Temporary placement of rotating bracket and locking saddle

Creating indentations with awl and hammer

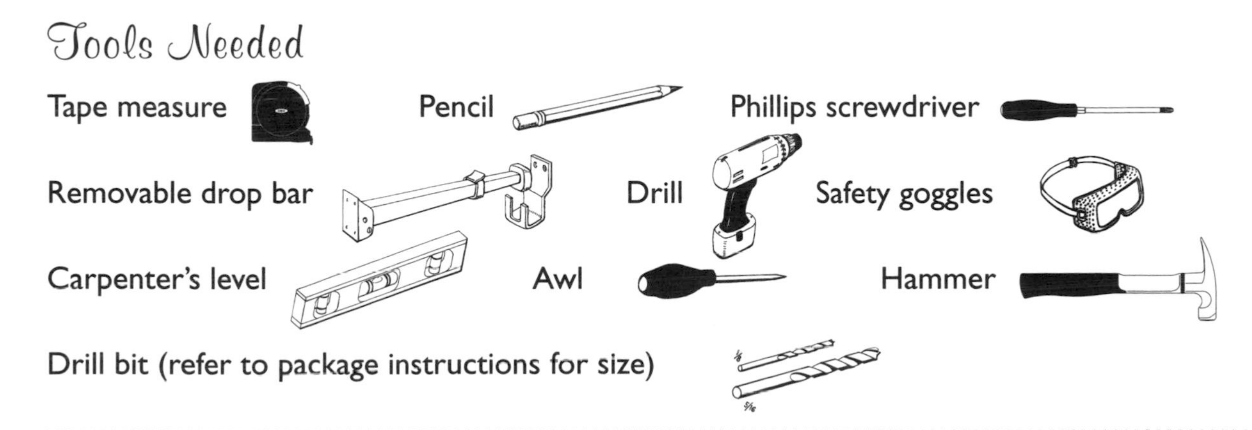

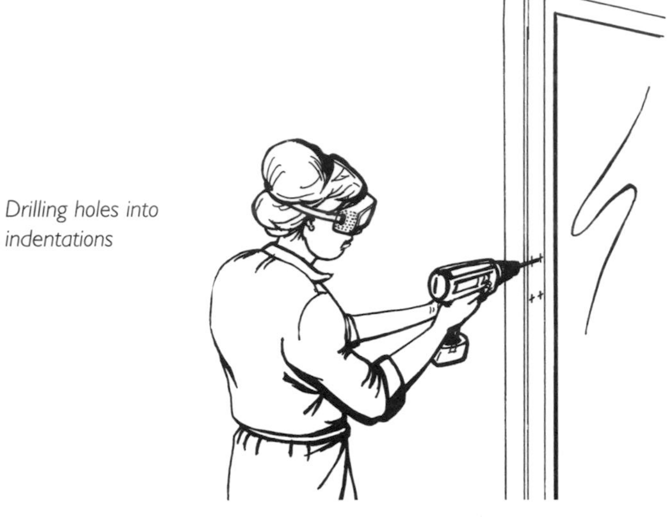

Drilling holes into indentations

Installing the Bar

Line up the bar so that the holes in the bracket and saddle are matched up with the holes you just drilled. Use the Phillips screwdriver to insert the screws and tighten.

Track Screws

Upper track screws are inserted into the top track of the sliding door to prevent someone from lifting the door out of its track. You *don't* want to insert the screws flush with the upper inside track, because in order to prevent a burglar from popping the door out, the screws have to protrude from the track.

Tape measure 

Pencil 

1/8" drill bit

Drill 

Safety goggles 

Three 1" or 1 1/2" #10 self-tapping screws 

Phillips screwdriver 

Preparing the Door

Completely open the sliding door. Measure the length of the upper inside track. Divide the length into thirds and mark with a pencil the location for three screws.

Drilling the Holes

Insert the drill bit into the drill just enough so that the chuck (the round part that houses the drill bit) securely grasps it. Put the safety goggles on and drill at the first mark until the bit passes completely through the track. Repeat this procedure at the other two marks. Remove the goggles.

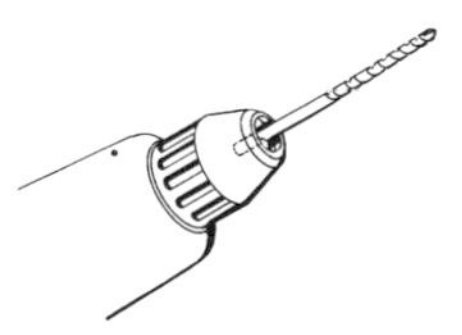

Inserting drill bit

Installing the Screws

Place a screw into a hole and tighten with the screwdriver, but do *not* make it flush with the track. Instead, insert far enough just to clear the door's travel. Repeat this procedure on the remaining two screws.

Close the door completely. If there is any resistance while closing the door, insert the screws a little deeper into the track.

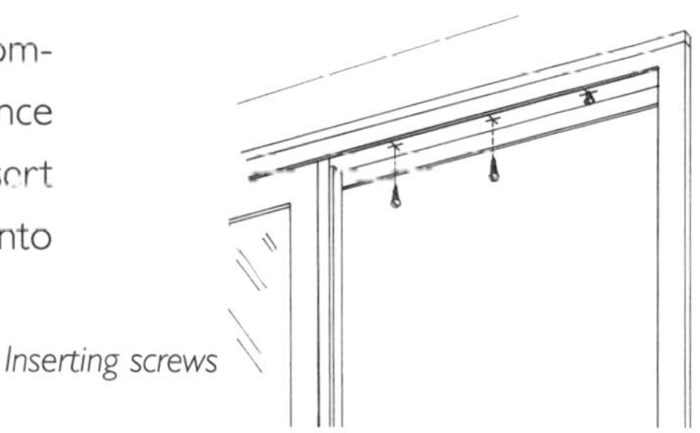

Inserting screws

Foot Lock

A foot lock is installed in the lower track where both doors meet. When installing it, you want to be careful not to drill into the door deeper than 1 ½ inches because you might break the glass. Before purchasing drill bits, find out if your door is wood or metal and read the manufacturer's instructions.

Some manufacturers recommend removing the sliding door during part of the installation. We, however, think that it can be done leaving the door in place. If in doubt, contact the manufacturer of your door.

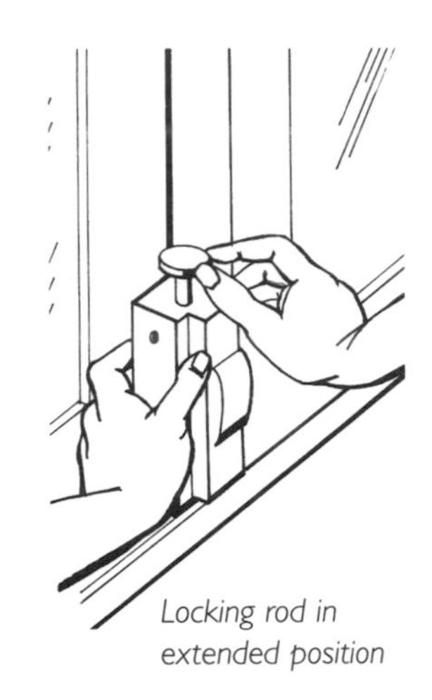
Locking rod in extended position

Preparing the Door

Close the sliding door and lock it. With the foot lock in one hand, place it flush with the front edge of the base of the sliding door. Place the locking rod in its extended position (i.e., unlocked) just high enough from the bottom so the rod will not interfere with the door's travel.

Drilling the Hole

Note: It's extremely important to keep the drill straight while drilling.

Creating indentations with awl and hammer

With a friend holding the lock in position, place the awl inside each of the lock's holes and tap with a hammer to make an indentation on the door's edge. (You're making a starter hole so that it will be easier to insert the drill bit.) Insert the drill bit inside the drill just far enough so that the chuck (the round part which houses the drill) securely

Tools Needed

Foot lock 

Helpful friend 

Awl 

Drill bits (refer to package instructions for size) 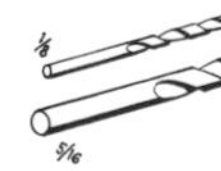

Hammer 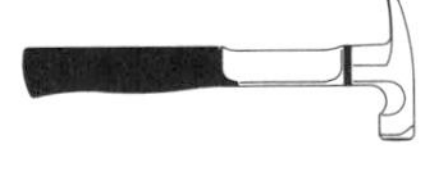

Drill 

Safety goggles 

Pencil 

grasps it. Put on the safety goggles and drill a starter hole ¼ inch deep through the indentations. Partially install the screws with the screwdriver.

Drilling hole into lower track

Hammer the top of the lock rod to make an indentation into the front lower track and mark the spot with a pencil. Now you have to remove the lock you so nicely installed. Using the starter drill bit, drill a hole into the lower track (on the marked spot) where the lock rod will be inserted. Replace the starter drill bit with the larger one and continue drilling until the hole is the correct diameter for the lock rod. The drilling can become a drag, and you may have to switch bits to enlarge the hole. Once drilling is completed, install the grommet (if applicable) into the hole.

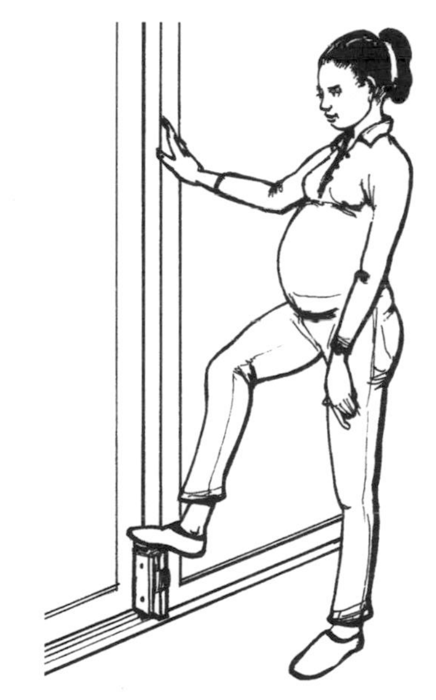
Engaging lock in track hole

Installing the Lock

Reinstall the lock into the door, this time permanently inserting the screws into the holes. Step on top of the lock to engage the rod into the track hole.

Resetting an Automatic Garage Door Opener

Let's face it, burglars are lazy. If they weren't, they'd be out there in the workplace like the rest of us. Don't make their job any easier! You can prevent burglars (or pranksters) from opening your garage door by changing the transmitter code.

Those of you who have a garage attached to your house should treat the connecting door as you would any other exterior door by keeping it locked. If you're going away for an extended period of time, unplug the garage door opener.

If your garage door opener was manufactured in 1997 or before, someone can use a device called a "code grabber" to open your garage door. The code grabber works by picking up codes being transmitted by any of the older garage door openers. All it takes is a point and click at your door and *open sesame*.

Some of the newer garage door openers have a non-repeating, rolling-code remote control that changes after every use. If you don't know which type of opener you have, consult the owner's manual or call the manufacturer. Also, check to see how often you should change the code.

Decoding the Opener

Place the ladder underneath the garage door opener (suspended from the garage ceiling) on the side farthest from the garage door. With the remote control in your pocket, climb high enough so you can safely and comfortably reach the unit.

Tools Needed

Ladder or stepladder

Garage door remote control

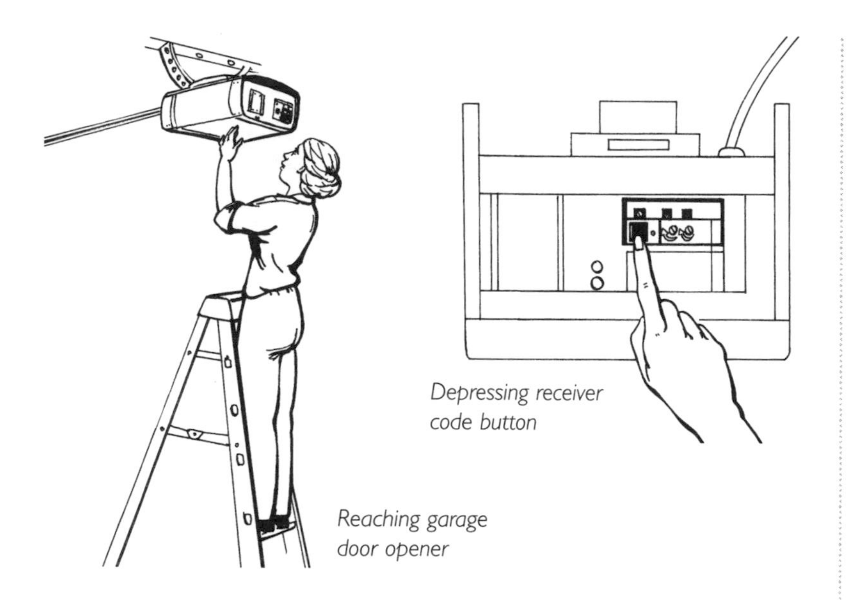

Depressing receiver code button

Reaching garage door opener

Typically, the unit (a.k.a. motor housing) has a receiver code button that programs the opener. Press and hold the receiver code button. An indicator light will come *on* to let you know the opener is erasing the code from its memory. When the light goes *off* (after about 6 seconds), the process is complete.

To test it, try to use the remote to open the garage door. If nothing happens, you succeeded in decoding the unit.

Resetting the Code

Press the transmitter button on the remote and hold it down. Push and release the receiver code button on the garage door opener unit. When the light flashes and you hear a clicking sound, the opener has been programmed correctly.

Make sure to test the opener by pressing the remote. If the garage door opens and closes, you've succeeded! If it doesn't, you may need to repeat the process or replace the remote control's battery.

Safety Measures

Performing Garage Door Safety Tests

Megan told us how she performed a safety test on her garage door the morning of her daughter's wedding—completely by accident. Her husband and daughter were waiting in the car, and to save time she decided to run under the garage door while it was closing. The door hit Megan (and then reversed), leaving a large black mark across the back of her new dress. Megan's story ended happily—the family arrived at church on time, the service was beautiful, and the only bruising was to her ego.

The garage door is the largest moving object in your home and should be considered lethal. Between 1982 and 1992, 54 children were killed by garage doors. In response to the danger that garage doors can pose, Congress passed legislation that mandates garage door openers manufactured after January 1, 1993, be required to have one of the following safety features: 1) control button that has to be held constantly to make the door close; 2) electric-eye sensor; or 3) door edge sensor (similar to those on elevators). All three features will stop the garage door before it ever touches an object. If your garage door was manufactured before 1982, you should be aware that some pre-1982 openers will stop when they hit an object, but they are not equipped with a reversal feature. The Consumer Product Safety Commission states that these openers cannot be adjusted or repaired to provide the automatic reversal feature found on later devices. Therefore, you should purchase a new garage door opener if it is more than 20 years old.

Mechanics of the Garage Door System

There are three parts to a garage door system: 1) door; 2) suspension; and 3) balance. You know what a door is, so we'll move right along to the suspension, which consists of the roller, tracks, pivot arms, and brackets. Balance pertains to the springs, cables, and pulleys.

Testing Your Garage Door System

You should perform three safety tests each month to ensure that your garage door is functioning properly: 1) door balance; 2) automatic reversal feature; and 3) force setting. (Only the balance test is required for a manual garage door.)

If a problem occurs during any of these tests, disconnect the automatic door opener until it can be repaired by a qualified service representative.

Always keep your garage door owner's manual handy, and if you don't have one, request a copy from the manufacturer.

Tools Needed

Large roll of paper towels (with the plastic wrapping intact)

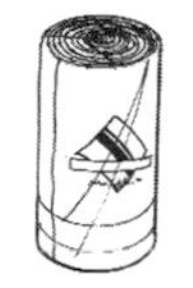

Ladder or stepladder

Small flathead screwdriver

Owner's manual

A manual garage door can be opened by hand, while an electric garage door can be opened using a designated wall switch, an electric garage door opener, or by disengaging the door and opening by hand.

Safety Test #1: Testing the Balance of a Manual or Electric Garage Door

Close the garage door. If your door is electric, disconnect the opener release mechanism by pulling on the handle attached to the string hanging from the trolley (also called carriage). This will allow you to operate the door by hand. Manually lift the door 3 to 4 feet above the ground and quickly move away. The door should move with little resistance and stay open. If the garage door drops to the ground, then it's out of balance. Reconnect the opener release mechanism only if the garage door passed this test. If it didn't, operate the door manually (refer to the owner's manual, if necessary) until it can be repaired by a qualified service representative.

Door lifted halfway above ground

Safety Test #2: Testing the Reversal Feature of an Electric Garage Door

Using the automatic door opener or wall switch, open the garage door. Place the roll of paper towels on the ground in the middle of the garage doorway (the roll of paper towels best simulates the density of a child's chest). Step away and push the control button to close the door. The door should reverse within 2 seconds of hitting the roll of paper towels.

If this doesn't happen, place the ladder underneath the motor

Testing reversing feature on electric garage door

housing and adjust the automatic reversal knob with your fingers or the screwdriver (refer to the owner's manual, if necessary). Repeat the steps until the door reverses correctly. If you can't correct the problem, release the garage door opener mechanism and operate the door manually until it can be repaired by a qualified service representative.

Safety Test #3:
Testing the Force Setting of an Electric Garage Door

Begin with the garage door completely open. Push the automatic garage door opener at the wall switch to engage the door to close. Hold up the bottom of the door as it closes (about halfway from the ground). The door should reverse immediately. If it doesn't, move the ladder underneath the motor housing. Decrease the down force by turning the adjustment control knob counterclockwise approximately 10 degrees with your fingers or the screwdriver (refer to the owner's manual, if necessary). Run the garage door through a complete open and close cycle before performing this test again, waiting 15 minutes between each cycle to prevent the motor from burning out. If you cannot correct the problem, disconnect the garage door opener mechanism and operate the door manually until it can be repaired by a qualified service representative.

Adjusting the force setting

Adjustment controls on motor housing

Adding Safety Devices to Window Treatments

Martha thought her home was the Fort Knox of baby-proofing: the safety plugs were in every outlet; the medicine cabinet was locked; and window guards were installed in all of the second-floor windows. At least that's what she thought until she saw her young child playing with the window cord.

Safety devices for window treatments are the most overlooked protective measure for children. Typically a window treatment has a long, looped cord, which when pulled, raises or lowers it. The Consumer Product Safety Commission states that between 1981 and 1991, 119 children strangled in the loop end of a window treatment cord.

Window safety products (e.g., safety tassels, cord stops, or cord cleats) are designed to allow consumers to use their window treatments without using a looped cord, and are available to the public *free* through the Window Covering Safety Council. These gadgets can also be purchased in stores that sell baby products or through safety catalogues.

Keep cribs, beds, and furniture away from windows.

Tools Needed

Scissors 

Safety tassels 

Cord stops 

Tape measure 

Pencil (sharpened) 

Cleats (includes screws) 

Phillips screwdriver 

Installing Safety Tassels and Cord Stops

Lower the blind to its maximum length and lock it in place by pulling the cord slightly to the side. Cut the looped cord above the tassel. Remove the old tassels and equalizer buckle, if applicable. Push the end of the cord into the hole in the smaller side of the new tassel and pull it through. Tie a few knots at the end to keep the tassel secure. Repeat this procedure on the other cord.

Holding one cord, make a loop 1 to 2 inches below the head rail of the blind. Push the end of the loop into the hole of the cord stop and pull it through. Take the end of the cord and insert it into the loop to create a knot. Tighten the knot by pulling on the cord. Repeat this procedure on the other cord.

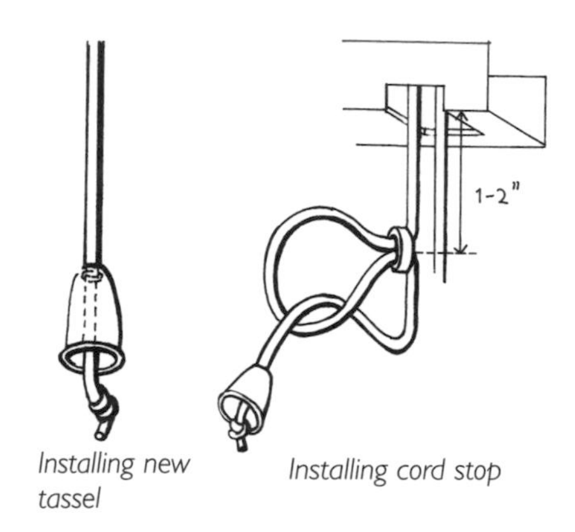

Installing new tassel

Installing cord stop

Installing Cleats

Cleats can be installed on the side of the window frame or on the nearby wall. Measure 6 to 12 inches between the two cleats and mark with a pencil. Make sure that the cleats are placed high enough to be out of a child's reach. Insert a screw into a cleat and, using the Phillips screwdriver, fasten it into the mark on the wall or on the side of the window frame. Repeat the procedure with the other cleat and screw. Wrap the cord around the cleats.

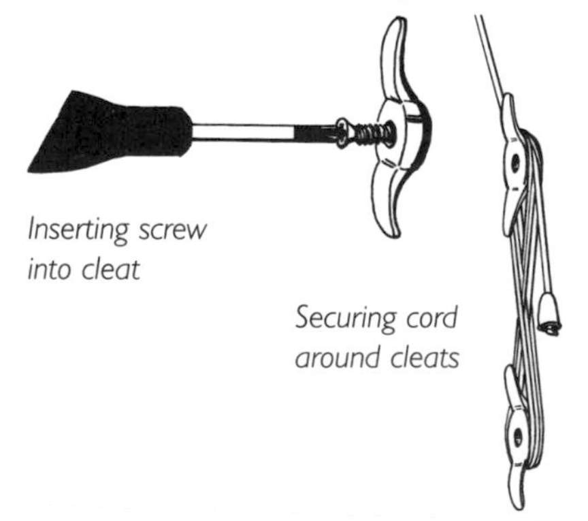

Inserting screw into cleat

Securing cord around cleats

Installing Interior Window Guards

T*ragic.* That's what we all said when we learned of the death of Eric Clapton's child in 1991. The fatal fall of the legendary musician's son from a fifty-third-floor window in New York City made news around the world. But what we didn't hear about were the thousands of other children who were either injured or killed that same year by falling out of a window. Since 1981, 140 children have died and approximately 4,700 children are injured each year by window falls. Most are under the age of five.

You can prevent this from happening in your home by following the window safety checklist *and* installing window guards. Those of you who live in an apartment should know that some cities require landlords to install safety guards on all windows (except those that lead to a fire escape). Contact your city government for more information.

If you're concerned that installing a window guard means the window always has to be shut, keeping you from exiting in case of a fire, you're wrong on both counts. Today's window guards are simple to install, provide easy removal in case of an emergency, and allow you to enjoy an open window.

Window Safety Checklist

- Keep furniture away from windows.
- If possible, open windows from the top rather than the bottom.
- A window opened from the bottom should be opened only 4 inches or less.
- Window screens are *not* window guards—they're only meant to keep bugs out.
- Check to be sure window screens are properly repaired and installed.
- Teach your children not to play near windows and never to lean against air conditioning units and window screens.
- Install window guards with easy-to-release mechanisms on second-story windows and above, or if the first floor is over 12 feet high.
- Have every age-appropriate family member practice releasing the window guards in case of a fire.
- Contact your local fire department to find out your town's requirements regarding window guards.

Purchasing a Window Guard

Before purchasing a window guard, measure the interior dimensions of the window frame (horizontal and vertical). Remember when measuring that not all windows in a home are the same size. Custom sizes can be requested. Window guards can be purchased through safety catalogues or on the Internet. Products and instructions may vary according to manufacturers.

Preparing the Window Guard

To adjust the guard to the correct size of the window, depress the two buttons located on the top grid. Widen or shorten as necessary. Snap the retractable grid securely into the side posts of the window guard. Position the window guard into the window frame, making sure to leave enough space between the guard and the window so that it doesn't interfere with the window's movement.

Installing the Window Guard

On one side, place the awl into one of the holes in a side post. Hammer the awl to make a starter hole, being careful not to pound the awl too far into the frame—its purpose is to create a starter hole to make it easier for you to insert the screw. Insert the screw through the side posthole and into the new starter hole in the frame. Tighten with the screwdriver. While repeating this process with the remaining screws, check to be sure that the window guard is level by eyeballing it.

Positioning window guard

Creating indentations with awl and hammer

Never open the window past the top of the side posts.

Tools Needed

Window guard (includes screws) 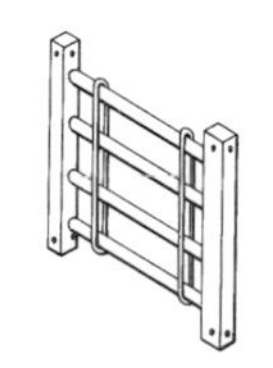

Awl

Hammer 

Phillips screwdriver 

Freeing Someone Locked Inside a Room

Jill's mother, who suffered from Alzheimer's, had a bad habit of locking herself in a room. At first it was just a mild inconvenience for Jill. But when it turned into an everyday problem, she decided it was time for "out with the old and in with the new"—not her mother, silly—the doorknobs! Jill removed all of the 40-year-old doorknobs, donated them to a local housing charity, then replaced them with new privacy doorknobs that have a lock-releasing mechanism.

Every privacy doorknob has a feature that enables you to unlock it from inside the room, as well as unlock it from outside, without a key. This feature can be a small hole in the center of the knob, a slot located in the middle of the knob, or a small hole found on the cover at the base of the knob (also called rosette). Depending on the make of your doorknobs, the tools and time it takes to unlock the door will vary.

If you find yourself unlocking the door too often, we recommend you keep one of the necessary tools on top of the doorframe or replace the doorknobs.

Tools Needed

Paper clip (straightened), bobby pin, or long metal nail 

Metal fingernail file or flathead screwdriver 

Knobs with a Hole

Push a straightened paper clip into the hole. When you hear it pop, you've unlocked the door.

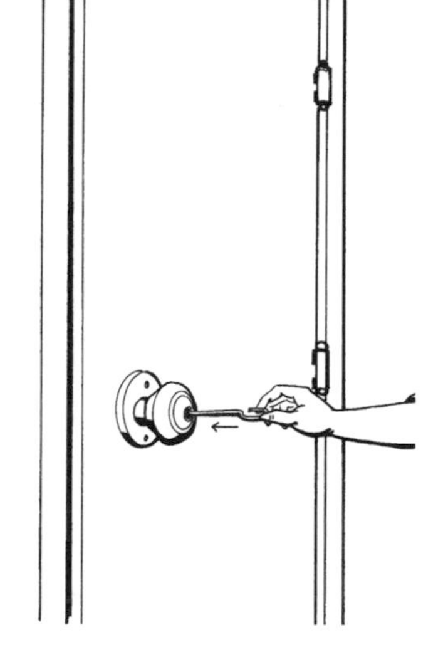

Releasing lock using straightened paper clip

Knobs with a Slot

Push the metal fingernail file or the flathead screwdriver into the slot. When you hear it pop, you've unlocked the door.

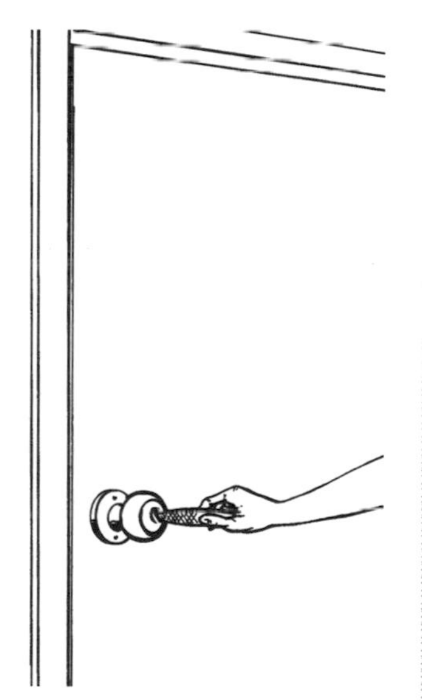

Releasing lock using metal fingernail file

Setting a Water Heater's Temperature

For the past 22 of her 78 years, Laurel has been a widow living alone on a fixed income. She was blessed with neighbors who were always willing to fix things around her house, but she began to feel that it was time to dare to tackle some of the projects herself. Laurel decided to start small by changing the temperature of her water heater. With a few successes under her tool belt, she began visiting her widowed friends to show them how to do small home repairs. "Who says you can't teach an old dog new tricks!" boasts Laurel.

Do you know how hot your water temperature is? Water temperature that is 140°F can produce first-degree burns within 3 seconds. You can ensure your water will be at a safe temperature by adjusting the setting of the water heater to 120°F. If you live in an apartment, have your superintendent install an anti-scalding device on your faucets. (Most new water heaters are pre-set to 120°F before shipment by the manufacturer.)

Tools Needed

Meat or candy thermometer 

Watch 

Testing the Water's Temperature

Before you make any adjustments to your water heater, first take your water's temperature.

Run the hot water in your bathtub or sink for approximately 5 minutes. With the hot water still running, hold the thermometer underneath for about 1 minute. Turn *off* the water and read the temperature on the thermometer. If the temperature reads above 120°F, you'll need to change the setting of the hot water heater. After you've adjusted the setting, wait 1 hour before retesting the water temperature.

Thermometer under running water

Running the faucet for 5 minutes wastes a lot of water, so perform this test after taking a shower, when the water is already hot!

Changing the Setting of a Gas Water Heater

A gas water heater has a temperature control knob located near the bottom that is used to adjust water temperature. Some knobs have words such as "vacation," "energy saving," "warm," "hot," or "hotter," or they have markings or indentations with temperature readings of 120°F, 130°F, 140°F, and 150°F. Some knobs will have both words and markings.

Before beginning, read the instructions located on the large label on the water heater or in the owner's manual.

Locate the thermostat control knob near the bottom of the water heater. If the knob clearly states 120°F, then turn it to that marking. If you are unsure as to which word denotes 120°F, check the owner's manual or contact the manufacturer.

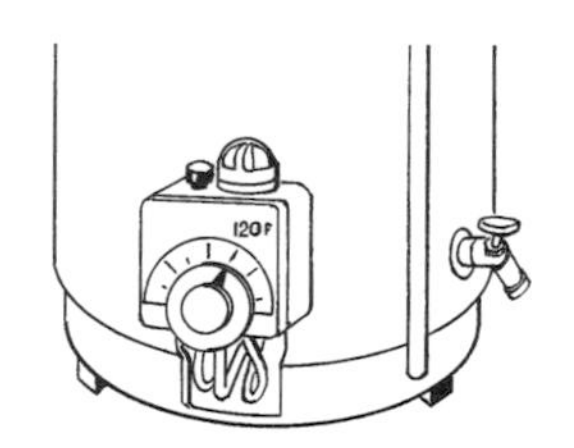

Temperature dial on water heater

Tool Needed

Owner's manual

Changing the Setting of an Electric Water Heater

An electric water heater has either one or two panel doors on the front with the thermostats behind them. The thermostat control knob allows you to adjust the temperature of the water heater and contains markings or indentations with temperature readings of 120°F, 130°F, 140°F, and 150°F. Some models require you to adjust one thermostat and others require you to adjust both. If you're unsure, check the owner's manual or contact the manufacturer.

Before beginning, read the instructions located on the large label on the water heater or in the owner's manual.

Turn *off* the power to the water heater at the main service panel. Remove the panel door(s) and move the insulation away from the thermostat. Do not remove the plastic cover.

Moving aside insulation

Use the screwdriver to adjust the thermostat to 120°F. Depress the reset button (typically red). Replace the insulation, making sure that the thermostat is well covered, and replace the panel door(s). Restore the power to the water heater.

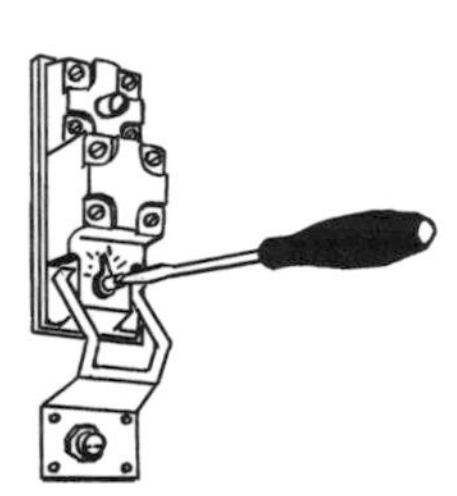

Adjusting thermostat

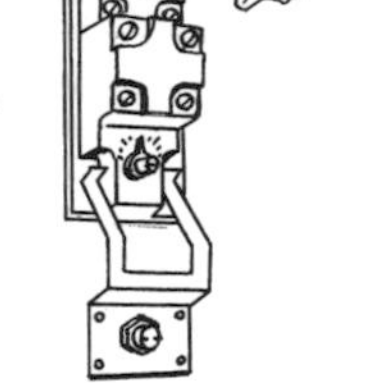

Depressing reset button

Tools Needed

Owner's manual

Flathead screwdriver

Installing a Handheld Shower Unit

Dottie's mother, Evangelia, was visiting for an extended period of time, so she decided to install a handheld shower unit in the guest bathroom. Dottie knew that it would make showering easier for her mom, but what she didn't know was how much nicer it would make her life, too. Once her kids started using the new showerhead, they fought over who would be first to take a shower. Who knows, showers today, tooth brushing tomorrow?

There are a lot of styles of handheld shower units, all with different instructions. Therefore, it's important to refer to the owner's manual throughout this installation. For example, some models come with washers (a.k.a. gaskets) and others are washerless.

The following instructions explain how to install a handheld shower unit with a washer and a vacuum breaker.

The majority of cities throughout the country require a handheld shower unit to be installed with a vacuum breaker, which keeps the backflow of water out of the community's water system, but not all units come with one. Check to see if the new unit comes with a vacuum breaker, or where you can purchase one.

Tools Needed

Towel 

Masking tape 

Adjustable wrench 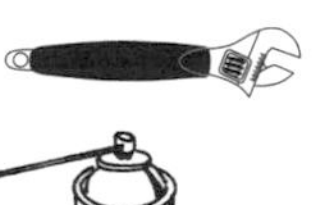

Slip-joint pliers or plumber's wrench 

Lubricating spray

Rubber grip or wash cloth 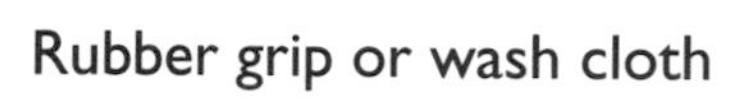 

Teflon tape 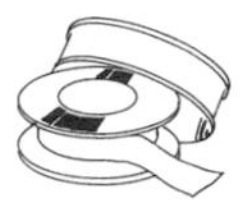

Old toothbrush or wire brush 

New handheld shower unit

Place a towel over the bathtub drain to keep any small parts from falling down the drain.

Removing the Old Shower Head

Wrap masking tape around the heads of the slip-joint pliers and adjustable wrench to protect the shower arm from being marred. Attach the slip-joint pliers to the top of the shower arm to keep it from rotating while you're trying to remove the head.

Position the rubber grip or washcloth on the collar nut and place the adjustable wrench on top of it, turning it counterclockwise to loosen. If the nut is difficult to unscrew, apply lubricating spray to it. Once the nut is loosened, you can continue unscrewing it by hand. Use a toothbrush to clean off any residue found on the threads.

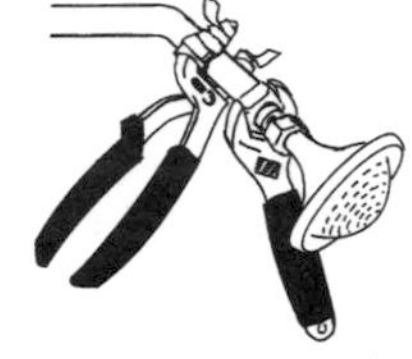

Slip-joint pliers and adjustable wrench on shower arm

Apply the Teflon tape to the exterior threads of the arm, wrapping it counterclockwise until you have three or four layers. Be sure to stretch the tape to get it into the grooves.

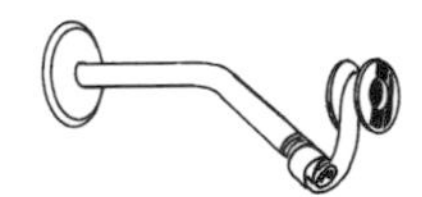

Applying Teflon tape to threads of shower arm

Installing a Handheld Shower Unit

Insert the vacuum breaker into the nut at the end of the hose, which attaches to the shower arm (if you install the vacuum breaker into the end of the hose that attaches to the *hand shower,* it will not work properly). Place the washer (if provided) firmly into the nut of the hose.

Place the hose on the end of the shower arm, turning it clockwise by hand until it's tightened. Insert the hand shower into the shower arm mount, rotating it to the desired position. Turn the shower *on* and look for leaks at the connection. It's common for a hand shower with a vacuum breaker to drain for 1 or 2 seconds after the shower has been turned *off.* If there is a leak at the connection, use the adjustable wrench to tighten the nut.

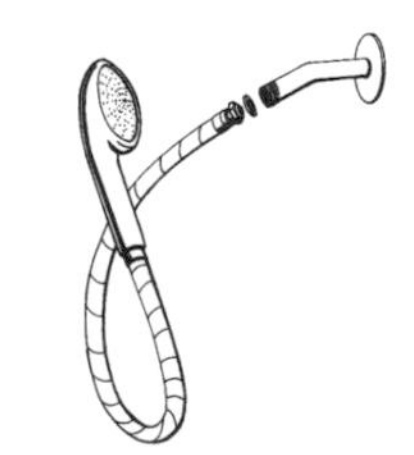

Attaching hose to shower arm

Replacing a Round Doorknob with a Lever

Virginia was *clueless in Seattle* about solving her doorknob problem. With her arthritis worsening, opening a door was no longer a mindless task—it had become a constant chore. She refused to be a prisoner in her own home, so she did some research and found that levers are the preferred hardware for people with arthritis. For Virginia, the project was simple, and it opened the door for more do-it-*herself* repairs.

When purchasing a lever, it's important to know the function of the door (see "Replacing a Doorknob," page 163), and its direction (i.e., whether the door swings right or left). The way to tell its direction is to stand outside a room facing the door. If the hinges are on the right, then it's a right-handed door; if the hinges are on the left, then it's a left-handed door.

If an adjustment needs to be made to the latch bolt (i.e., the tubular part of the latch located in the middle of the door hole), check the manufacturer's instructions for easy realignment.

Tools Needed

Phillips screwdriver 

New door lever 

Removing an Old Doorknob

Use the Phillips screwdriver to loosen the two screws on the cover at the base of the doorknob. The best way to do this is to hold down both covers (one on each side of the door) with one hand, while using the other hand to remove the screws.

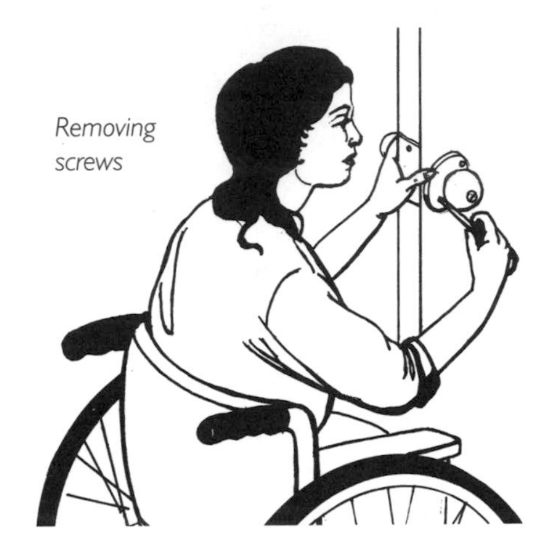
Removing screws

Installing a Lever

Be sure to place the hardware containing the lock on the interior side of the door. Insert the stems of the exterior lever horizontally into the holes in the latch case. Place the interior lever on the protruding spindle, carefully aligning the stems with the screw holes.

Hand turn the screws into the holes, while holding both covers with the other hand. Use the Phillips screwdriver to tighten the screw closest to the door edge first. Tighten the other screw, and test the lever by opening and closing the door.

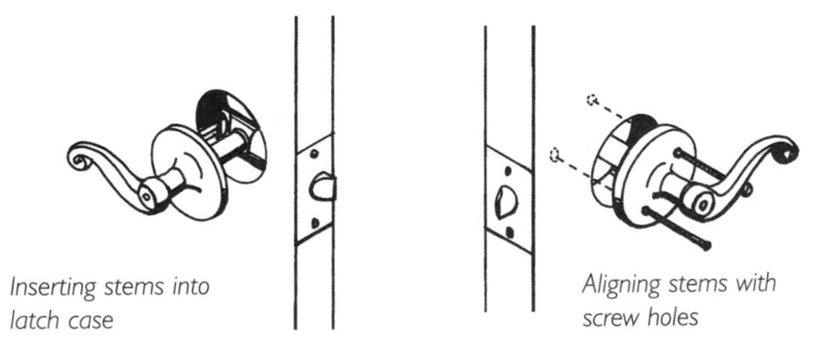
Inserting stems into latch case

Aligning stems with screw holes

Preventing Falls

When the final house had been sold, the Bridge foursome (a.k.a. the Golden Girls) headed from New York, where they had lived all their lives, to Florida. Good-bye to snow, hello to warm weather, and who knew, maybe some Sunshine Boys? But first things first. None of them was a spring chicken anymore, so they had to be realistic about setting up house. They contacted the AARP about how to make their home senior-friendly (i.e., accident-proof). After contracting out some items and doing the others themselves, the Golden Girls were playing their cards just right.

Falls are the leading cause of accidents for the elderly in their homes. Although we can't guarantee that you'll never fall in your home, there are some easy ways to create a safer environment, such as securing rugs to the floor and applying colored tape to stairs and landings. Having extra lighting and handrails professionally installed is also a great safeguard against falls.

Tools Needed

Scissors 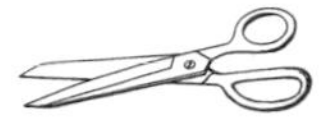 Non-skid tape  Colored or reflective tape 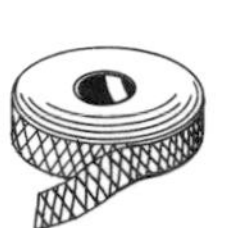

Securing a Rug

Applying non-skid tape to rug

All rugs (including doormats) should be secured to the floor with non-skid tape. This special double-sided tape can be purchased in carpet or hardware stores.

Before beginning, make sure that the floor and the bottom of the rug are clean and dry. Turn the rug over so that the underside is facing up. Cut the tape to the desired size and secure it liner side up to the four sides of the rug and in the middle. Remove the liner. Flip the rug over and return it to its location, pressing it down.

Applying Contrasting Color Strips to Stairs, Landings, and Step-downs

Applying tape to stairs

Stairs cause the second greatest number of accidents for seniors, so for those with visual impairments, it's vital to place colored strips on stairs, landings, and step-downs. Red and yellow-orange are the colors easiest to discern. Reflective tape can also be helpful, especially in the dark.

Cut the tape to the correct size. Secure the tape to either the first and last stairs to distinguish a change in level, or to all of the stair treads. Step-downs should also be marked with colored or reflective tape.

Practicing Ladder Safety

At 5 feet 10 inches, Donna was a walking stepladder, but even she couldn't reach everything in her home. Never wanting to take the time to look for the stepladder, Donna would instead find creative substitutes: a bureau, kitchen countertop, and all types of chairs. It wasn't until she had a bad fall that she was brought down to size.

We're all guilty of using a chair rather than a stepladder to reach things in our house. Of course, there are some of you who are a bit more creative (and crazier) than we, who have climbed on top of bureaus and kitchen countertops to reach that unreachable object.

We want you to get smart before you get hurt, so buy a stepladder and store it in a convenient place. Oh, and one more thing . . . repeat after us: *A chair is for sitting, not for standing.*

There are four types of ladders most commonly used around the home: 1) stepladder; 2) straight; 3) extension; and 4) sectional. A stepladder, the shortest of the ladders, has two or three steps, and a spreader on each side that locks into place. A straight ladder comes in heights ranging from 4 to 12 feet. An extension ladder is typically 16 feet but can extend to 40 feet. A sectional ladder transforms into an A-frame or extension ladder.

Stepladder

Straight ladder

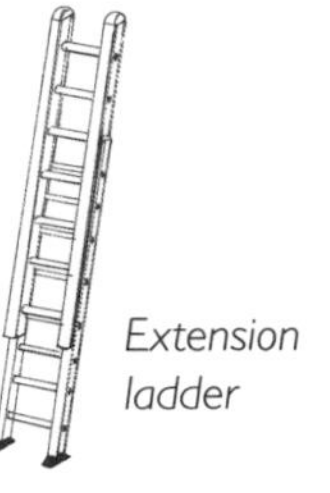

Extension ladder

Sectional ladder

Ladders are rated according to their weight capacity. Household ladders should be rated either Type III (200 pounds maximum) or Type II (200–250 pounds maximum). Look for the rating sticker on the side of the ladder.

Safety Rules

Once you've chosen the right ladder for the job, adhere to the following rules:

- **Never use a damaged or worn ladder.**
- **Determine the proper angle at which to place the ladder.**
- **Don't place a ladder near overhead electrical wires.**
- **Wear a tool belt to carry supplies.**
- **Don't go up a ladder if the ground is wet or it's a windy day.**
- **Never climb higher than the third rung from the top.**
- **Always keep your hips between the side rails with your legs spread slightly for balance.**
- **Never have more than one person on the ladder.**
- **Carry the ladder parallel to the ground.**
- **Check to be sure that the ladder is level before climbing.**

Tools Needed

Helpful friend (if necessary)

Ladder

Tape measure

- Wear shoes with good traction.
- Never overreach.
- Always lock both spreaders on a stepladder.
- Hold on to the ladder with at least one hand.
- Never stand on the top of a stepladder.

Ladder Safety Formula

To determine the correct angle at which to place the ladder against the exterior wall, use this formula: Length of ladder/4= Correct Angle. For example, if the length of the ladder you are using is 16 feet, then the equation will be 16 feet/4 = 4 feet. Therefore, the distance between the ladder and the wall should be 4 feet.

Ladder Placement

With a helpful friend, carry the ladder horizontally to the desired location, and place it against the exterior wall. Measure the distance between the base of the ladder and the exterior wall. If necessary, adjust the placement of the ladder until the correct measurement is achieved.

Proper ladder placement

Patching Cracks in a Driveway

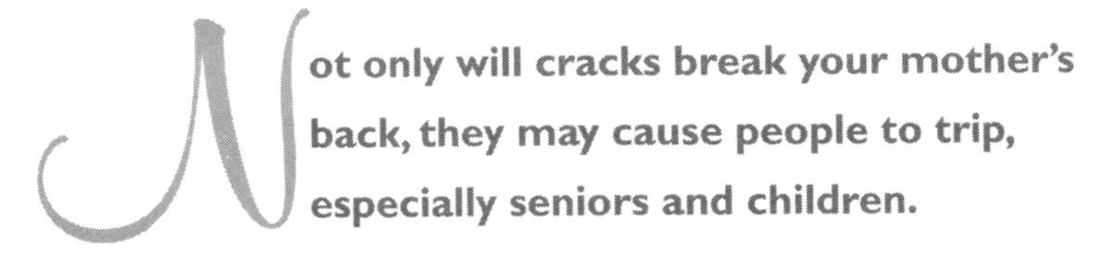

Not only will cracks break your mother's back, they may cause people to trip, especially seniors and children.

Asphalt and concrete sealants are an inexpensive and easy way of solving the problem. New technology has brought about a more flexible and durable sealant that can easily be installed in varying temperatures. However, manufacturers suggest that you do this repair only when you can be assured it won't rain for 24 hours afterward. As always, read the manufacturer's instructions before beginning.

Repairing Asphalt Driveways

Preparing the Surface

Use the push broom to sweep away any dirt and pebbles from the crack. Measure the depth and width of the crack: if it is less than $\frac{3}{8}$

Tools Needed

Push broom

Tape measure 

Asphalt or concrete sealant (tube) 

Polyethylene foam (available in different widths) 

Flathead screwdriver 

Scissors or utility knife 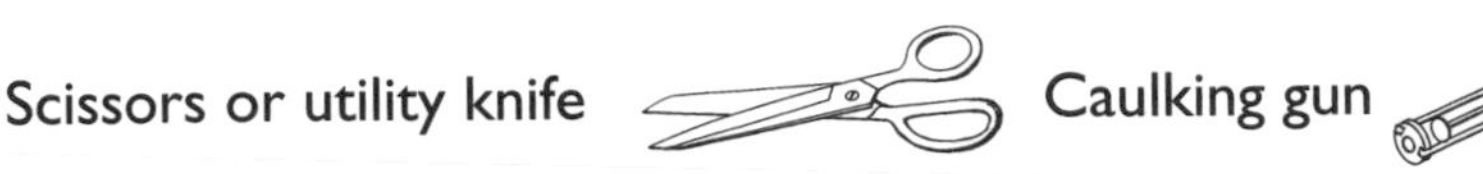

Caulking gun 

inch in depth or width, then you can proceed just by filling the crack with the sealant.

For cracks that measure more than 3⁄8 inch in depth or width, purchase foam that is at least 1⁄8 inch thicker than the crack for a tight seal.

Filling the Crack

Cut a piece of foam to the appropriate length and push the foam into the crack with your fingers or the screwdriver. (It's not necessary to fill the entire crack, just the upper portion.) Leave a maximum space of 3⁄8 inch at the top of the crack for the sealant.

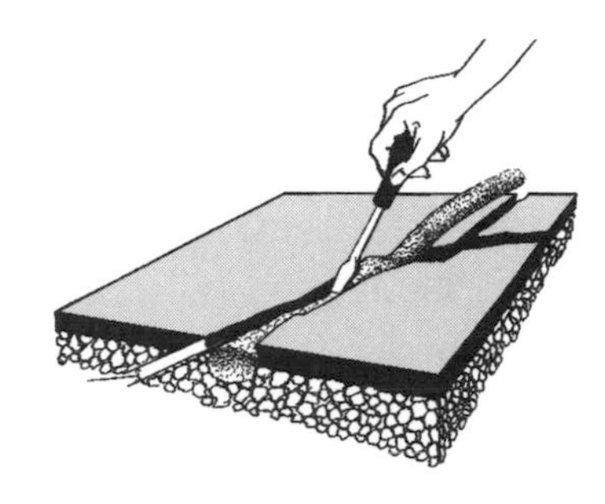

Inserting foam into crack

Applying the Sealant

Take the cap off the sealant and use the blade and the seal puncture on the caulking gun to snip off the top of the tube and create a hole in the opening. Load the tube into the caulking gun and squeeze the handle. Apply the sealant evenly over the foam, and replace the cap. Allow the sealant to set for 24 hours.

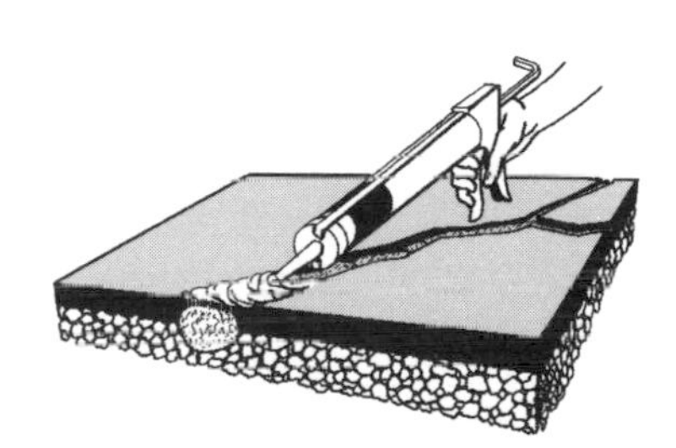

Applying sealant

Repairing Concrete Driveways

Preparing the Surface

Use the push broom to sweep away any dirt and pebbles from the crack. Measure the depth and width of the crack with the tape measure. If the crack is less than ½ inch in depth or width, you can proceed by filling the crack with the sealant.

For cracks that measure more than ½ inch in depth or width, purchase foam that is at least ⅛ inch thicker than the crack for a tight seal.

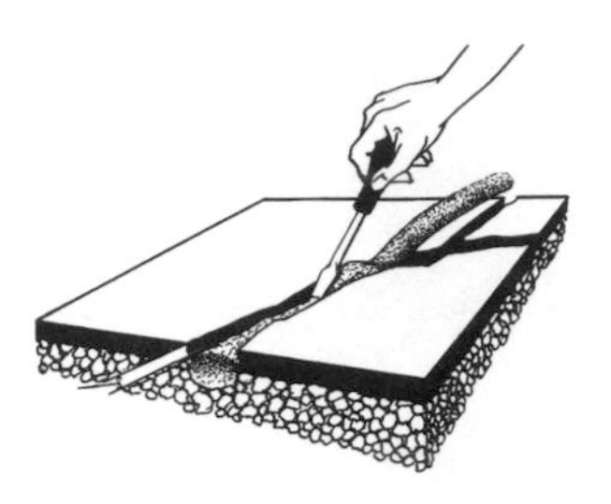

Inserting foam into crack

Filling the Crack

Cut a piece of foam to the appropriate length and push the foam into the crack with your fingers or the screwdriver. (It's not necessary to fill the entire crack, just the upper portion.) Leave a maximum space of ¼ inch at the top of the crack for the sealant.

Applying sealant

Applying the Sealant

Take the cap off the sealant and use the blade and the seal puncture to snip off the top of the tube and create a hole in the opening. Load the tube into the caulking gun and squeeze the handle. Apply the sealant evenly over the foam, and replace the cap. Allow the sealant to set for 24 hours.

Safety Checklist for Children and Seniors

We've provided a checklist for making a home safe for children and seniors. The safety measures featured in this section were chosen based on the level of ease. For more information on this subject, contact the American Association of Retired Persons (AARP), the National Institute on Aging, the National Safety Council, and the Consumer Product Safety Commission.

Child Safety

- Use safety gates at the top and bottom of stairs.
- Insert child-proof plastic guards in receptacles.
- Install safety latches on all cabinets where medicines or hazardous materials are stored.
- Place oversize protectors on stove knobs.
- Install doorknobs inside closets.
- Use pots and pans only on the back burners.
- Keep appliance cords short by using cord shorteners.
- Never leave a child unattended in a bathtub or pool—an adult should always be present.
- Install safety tassels on all window treatments with cords.
- Don't use long phone cords.
- Apply colorful stickers to glass doors and non-skid appliqués to bathtubs.

- Unplug all countertop appliances when not in use.
- If you're disposing of a refrigerator or freezer, wrap the doors of the appliance shut with duct tape to prevent kids from playing inside.
- Install window guards on all windows except those leading to a fire escape.
- Set your water heater's temperature to 120°F.
- Install lock-releasing doorknobs.
- Plan an emergency escape route.
- Perform garage door safety tests.

Senior Safety

- Keep passageways clear.
- Move lightweight objects to upper shelves and heavier items to lower shelves.
- Secure rugs, including doormats, with non-skid tape.
- Add contrasting colored tape to first and last steps, landings, and step-downs to identify a change in level.
- Set your water heater's temperature to 120°F.
- Check for loose railings and flooring in decks and stairs.
- Plan an emergency escape route.
- Add a peephole to the front and back doors.
- Install lever faucets on sinks.
- Remove high thresholds in doorways to prevent tripping.
- Add light switches to the top and bottom of staircases and in closets.
- Install grab bars in bathtubs.
- Replace round doorknobs with levers.
- Perform garage door safety tests.

Households are changing: A 1999 United States Census Bureau report found that 3.9 million children are living in homes maintained by their grandparents. Another fast-growing segment of the population is women who have a young child and an elderly parent living under one roof—coined the "sandwich" effect. The safety measures taken for a child can at the same time be restrictive for a senior. For example, inserting safety plugs into receptacles is a great way of keeping kids' fingers out of sockets, but for an elderly person with arthritis, the safety plugs can be a formidable challenge. Installing a safety lock on a toilet will stop children from throwing things into it, but for the elderly, it's an unnecessary burden. The solution is to find a balance in your home between safety and usability.

Finding a Contractor

Now that we've taught you how to do basic home repairs, we're going to tell you how to hire a contractor for the projects that may be over your head, for now at least.

The biggest mistake people make when hiring a contractor is not doing the proper research ahead of time. The best thing you can do is to arm yourself with a short list of plumbing, electrical, and HVAC (Heating, Ventilation, and Air Conditioning) contractors. Think of it like finding a baby-sitter—you'll never find one if you wait till the last minute.

How do you acquire a short list? First start with your inner circle of neighbors and friends to ask for names of contractors they've used. If that doesn't work, then ask co-workers or acquaintances at your child's school. Once you have some contractors' names, call and ask the following and get the answers in writing:

- **Are estimates free?**
- **Is there a minimum charge or trip fee?**
- **Can I have a copy of your business license and insurance verification?**
- **Is the company bonded?**
- **How long has the company been in business?**

Another option is to pay a contractor referral service to do the background check for you. These companies can be found on the Internet or in the Yellow Pages.

Remember, if the offer sounds too good to be true, it probably is.

There are some red flags that signal trouble: A contractor saying it's a one-time offer; a quoted price that is *extremely cheap*; a post office box used as a business address; and a business license that the state says doesn't exist.

One other thing, don't bother asking a contractor for a list of referrals, because you'll only be given phone numbers of happy customers, not dissatisfied ones.

Resources

The following businesses, manufacturers, government agencies, and associations contributed their knowledge to *Dare to Repair*. It gives us great pleasure to thank them publicly and we hope that you'll contact them for information or products.

General

United States Consumer Product Safety Commission
4330 East West Highway
Bethesda, MD 20814
800-638-2772
www.cpsc.gov

National Safety Council
1121 Spring Lake Drive
Itasca, IL 60143
800-621-7619
www.nsc.org

Tools

Manufacturer

OXO International
75 Ninth Avenue, 5th Floor
New York, NY 10011
212-242-3333
www.oxo.com

Plumbing

Manufacturers

Delta Faucet Company
Attention: Product Service
55 East 111th Street
Indianapolis, IN 46280
800-345-3358
www.deltafaucet.com

Alsons Corporation
3010 West Mechanic Road
Hillsdale, MI 49242
800-400-5640
www.alsons.com

Fluid Master Inc.
30800 Rancho Viejo Road
San Juan Capistrano, CA 92675
800-631-2011
www.fluidmaster.com

Associations

Plumbing and Draining Institute
45 Bristol Drive
South Easton, MA 02375
800-589-8956
www.pdionline.org

Plumbing, Heating & Cooling Contractors National Association
180 South Washington Street
Falls Church, VA 22046
800-533-7694
www.phccweb.org

Ceramic Tile Institute of America, Inc.
12061 Jefferson Blvd.
Culver City, CA 90230
310-574-7800
www.ctioa.org

Electricity

Manufacturers

Greenlee Textron
4455 Boeing Drive
Rockford, IL 61109
800-435-0786
www.greenlee.textron.com

Associations

Electrical Safety Foundation International
1300 N. 17th Street, Suite 1847
Rosslyn, VA 22209
703-841-3229
www.electrical-safety.org

International Association of Electrical Inspectors
P.O. Box 830848
Richardson, TX 75083
800-786-4234
www.iaei.com

Major Appliances

Manufacturers

General Electric
3135 Easton Turnpike
Fairfield, CT 06431
800-626-2000
www.geappliances.com

Sears
3333 Beverly Road
Hoffman Estates, IL 60179
800-469-4663
www.sears.com

Whirlpool Corporation
2000 North M-63
Benton Harbour, MI 49022
800-253-1301
www.whirlpool.com

Maytag Corporation
403 West Fourth Street
Newton, IA 50208
800-688-9900
www.maytag.com

York International
5005 York Drive
Norman, OK 73069
405-364-4040
www.york.com

Associations

Gas Appliance Manufacturer's Association
2107 Wilson Boulevard, Suite 600
Arlington, VA 22201
703-525-7060
www.gamanet.org

Association of Home Appliance Manufacturers
1111 19th Street, NW, Suite 402
Washington, DC 20036
202-872-5955
www.aham.org

Windows, Walls, and Doors

Manufacturers

Andersen Windows
100 4th Avenue North
Bayport, MN 55003
888-888-7020
www.andersenwindows.com

Pella Corporation
102 Main Street
Pella, IA 50219
888-84-PELLA
www.pella.com

DAP Inc.
2400 Boston Street, Suite 200
Baltimore, MD 21224
888-DAP-TIPS
www.dap.com

Chamberlain Group
6020 South Country Club Road
Tuscon, AZ 85706
800-528-9131
www.chamberlain.com

Purity Max
880 Facet Road
Henderson, NC 27536
800-334-6659
www.purolatorair.com

Association

American Architectural Manufacturers Association
1827 Walden Office Square
Schaumburg, IL 60173
888-323-5664
www.aamanet.org

Home Safety

Manufacturers

M.A.G. Engineering & Manufacturing Company Inc.
15381 Assembly Lane
Huntington Beach, CA 92649
800-624-9942
www.magsecurity.com

First Alert, Inc.
3901 Liberty Street Road
Aurora, IL 60504
800-392-1395
www.Firstalert.com

Guardian Angel Window Guards
Automatic Specialties
422 Northboro Road
Marlboro, MA 01752
800-445-2370
www.auspin.com

Associations

National Fire Protection Association
1 Battery March Park
Quincy, MA 02269
617-770-3000
www.nfpa.org

AARP
601 E Street
Washington, DC 20049
202-434-2277
www.aarp.org

Chimney Safety Institute of America
800-536-0118
www.csia.org

Wood Floor Covering Association
2211 East Howell Avenue
Anaheim, CA 92806
800-624-6880
www.wfca.org

Window Covering Safety Council
355 Lexington Avenue
Suite 1700
New York, NY 10017
800-506-4636
www.windowcoverings.org

National Glass Association
8200 Greensboro Drive, Suite 302
McLean, VA 22102
866-342-5642
www.glass.org

Service Magic
1626 Cole Boulevard
Building 7, Suite 200
Golden, CO 80401
www.servicemagic.com

Women Helping Women

Here are just a few of the wonderful nonprofit organizations that help build or renovate homes for the disadvantaged, while revitalizing neighborhoods as well.

Habitat for Humanity International
Fiona Eastwood, Manager
Women Build Program
121 Habitat Street
Americus, GA 31709
800-HABITAT
www.habitat.org

Neighborhood Reinvestment Corporation
Ellen Lazar, Executive Director
1325 G Street, NW, Suite 800
Washington, DC 20005
202-220-2410
www.nw.org

McAuley Institute
Jo Anne Kane, Executive Director
8300 Colesville Road, Suite 310
Silver Spring, MD 20910
301-588-8110
www.mcauley.org

S.H.I.P. (Senior Home Improvement Program)
The Centre for Women
305 S. Hyde Park Avenue
Tampa, FL 33606
813-251-8437
www.centreforwomen.com

Rebuilding Together with Christmas in April
Patty Johnson, Executive Director and CEO
1536 16th Street, NW
Washington, DC 20036
800-4-REHAB-9
www.rebuildingtogether.org

TOP/WIN, Inc.
Linda Lyons Butler, President
2300 Alter Street
Philadelphia, PA 19146
Email: *topwin1@aol.com*

Index